DRILLS AND DRILL PRESSES

HOW TO CHOOSE, USE AND MAINTAIN THEM

RICK PETERS

Sterling Publishing Co., Inc.
New York

Acknowledgements

Butterick Media Production Staff

Design: Triad Design Group, Ltd.
Cover Design: Elizabeth Berry
Photography: Christopher J. Vendetta
Illustrations: Greg Kopfer, Bob Crimi
Cover Photo: Brian Kraus, Butterick Studios
Photo Supervisor: Tony O'Malley

Assoc. Art Director: Monica Gaige-Rosensweig
Copy Editor: Barbara McIntosh Webb
Page Layout: David Joinnides
Index: Nan Badgett
Assoc. Managing Editor: Stephanie Marracco, Nicole Pressly
Project Director: Caroline Politi
President: Art Joinnides

Special thanks to the following companies for providing product and technical support: DeWalt Industrial Tool Company, Makita USA, Milwaukee Electric Tool Company, Porter-Cable, S-B Power Tool Company, and Vermont-American Tool Corporation. I'm also grateful to Ernest J. Lamadrid of Arizona Power Tools Service for sharing his years of tool repair experience. Thanks to the production staff at Butterick Media for their continuing support. And finally, a heartfelt thanks to my constant inspiration: Cheryl, Lynne, Will, and Beth.   R.P.

Library of Congress Cataloging-in-Publication Data

Peters, Rick
Drills and drill presses : how to choose, use and maintain them / Rick Peters
p. cm.
ISBN 0–8069–3691–6
1. Power tools. I. Title.
TJ1263.P32 2000
621.9'52—dc21   99–086642

ISBN 0-8069-3691-6

Published by Sterling Publishing Company, Inc.
387 Park Avenue South, New York, N.Y. 10016
©2000, Butterick Company, Inc., Rick Peters
Distributed in Canada by Sterling Publishing, c/o Canadian Manda Group, One Atlantic Avenue, Suite 105, Toronto, Ontario, Canada M6K 3E7
Distributed in Great Britain and Europe by Cassell PLC, Wellington House, 125 Strand, London WC2R 0BB, England
Distributed in Australia by Capricorn Link (Australia) Pty. Ltd., P.O. Box 6651, Baulkham Hills, Business Centre, NSW 2153, Australia

Printed in Hong Kong
All rights reserved

B
THE BUTTERICK® PUBLISHING COMPANY
161 Avenue of the Americas
New York, N.Y. 10013

INTRODUCTION

A kid in a candy store. That's the way I felt when I bought my first portable drill. There I stood, hands and face pressed up against the display case, hot breath fogging the glass. I stared at the myriad choices, desperately wanting to make the right decision.

To this day, I remember the feeling. And although many years have passed since then, I can still identify with the glazed-over look on the faces of prospective power tool buyers that I bump into wandering through the aisles at the local hardware store or home center.

As a woodworker and homeowner with over 25 years' experience handling virtually every conceivable home improvement job (including the restoration of an 1890 Victorian home) and well over a decade in publishing (as an editor for *ShopNotes* and *Woodsmith* magazines and numerous book publishers), it still amazes me how little help there is out there for anyone looking to buy and use a power tool.

My purpose in writing this book is to put at your fingertips the information you'll need to select, use and maintain a portable drill or drill press—two of the most versatile power tools that you can own.

Whether you're looking for a light-duty, intermediate, or industrial-strength portable drill or drill press, Chapter 1 will help you identify the different types of portable drills and drill presses and the available features. To make the decision-making process less traumatic, I've provided decision-making flowcharts that should help you choose the non-impact drill, impact drill, or drill press that's right for you.

Chapter 2 describes the wide variety of accessories you can use to boost the performance of your portable drill or drill press, whether you're drilling in wood, metal, glass, plastic, or masonry; drill bits and specialty bits like countersink bits and hole saws, and accessories such as dowel jigs and vertical stands for portable drills, and circle cutters and rotary planers for the drill press.

In Chapters 3 and 4, there's a wealth of tips and techniques on using portable drills and drill presses for almost any job, as well as ways to increase accuracy and precision.

I've included some of my favorite shop-made jigs in Chapter 5—drilling guides for the portable drill, and an auxiliary fence and tables for the drill press. Simple and inexpensive to build, these jigs will expand the usefulness of your portable drill or drill press.

Chapter 6 delves into maintaining your portable drill or drill press. There's practical advice not only on how to keep these tools running in tip-top condition, but also in-depth how-to on how to repair common problems.

Repairs like replacing cords, brushes, and switches on portable drills. And adjustments like aligning belts and pulleys and eliminating runout on the drill press. There's also a section on cleaning, lubricating, and sharpening drill bits.

Finally, in Chapter 7, I'll discuss many of the solutions to the everyday problems that you're likely to face using your drill or drill press to tackle a wide variety of jobs—everything from wandering bits, crooked holes, rough holes, and burning to dealing with a portable drill or drill press that bogs down, catches, or stops.

All in all, this makes for a comprehensive guide that will help you get the most out of your drill or drill press. Armed with this information, I hope that you'll reach for your drill or drill press with greater confidence to handle an even wider variety of jobs around the home and shop.

Rick Peters
Spring 2000

CHAPTER

1 SELECTING A DRILL OR DRILL PRESS

The starting point for selecting a drill or drill press is similar to that for any tool, computer, or appliance: Ask yourself what you're planning on using it for. Anticipating future jobs you'll want to do isn't easy, but it's worth the effort—it's one of the best ways to prevent underbuying. To help you with this, I've broadly categorized jobs into three groups: light-duty, average-duty, and heavy-duty; see the chart *below* for some examples. If you find that most of your tasks fall into one category, you can use this with the decision-making charts on *pages 18 and 27* to help select the drill or drill press that's right for you.

Another way to define jobs is to consider whether a specific task requires sustained use or high torque. By sustained use, I mean jobs where the drill or drill press is running continuously for long periods. This includes removing house paint with a sanding disk, drum sanding on the drill press, wire-brushing a metal patio set in preparation for painting, or using a buffing pad to wax the car. High-torque applications could be driving screws (particularly in hardwoods), drilling large holes, cutting holes on the drill press with a circle cutter, or drilling in tough materials like masonry.

Once you've defined the type of work you'll be doing, the next step is to examine the numerous features available: configuration, power type, motor and battery ratings (for cordless drills), speeds and ranges, chuck capacity and type, table and tilting mechanisms (for drill presses), and ergonomics. To help steer you away from overbuying or underbuying, I'll share recommendations and tips about the wealth of choices on the market.

USAGE CHART

Light-Duty	Average-Duty	Heavy-Duty
• simple repair jobs, such as fixing a broken closet rod or shelf bracket • drilling holes for wall anchors to hang pictures, towel racks, etc. • basic installation work, like mounting curtain rods or adding a ceiling fan or light • light sanding or buffing	• advanced repair work, like replacing flashing or trim on the exterior of the house • more advanced installation jobs, like hanging drywall • basic woodworking, such as building a bookshelf out of softwood or plywood • removing paint or rust with a sanding disk or wire brush or wheel	• building a deck • remodeling a kitchen or bathroom • adding on to a room or installing a built-in such as a bookcase • advanced woodworking, using hardwoods and composite materials

Portable drills Portable drills are manufactured in more styles, shapes, and sizes than any other portable electric tool. Portable drills are categorized as being either impact or non-impact, and both types are available in corded or cordless versions.

Impact drills are further subdivided into hammer drills and rotary hammer drills (*see the sidebar on page 17* for more on this). Shown here *from left to right* are a ⅜"-chuck corded drill, a ⅜"-chuck cordless driver/drill, a ½"-chuck hammer drill, and a ½"-chuck rotary hammer drill.

Drill presses Drill presses come in three basic designs (*from right to left*): bench-top, floor-model, and radial drill press. The noticeable difference between a floor model and a bench-top drill press is their work capacity; that is, the distance between the drill chuck and the base.

Most floor models can accept a workpiece with a maximum length of 44" to 48". A bench-top drill press can usually only handle a workpiece with a maximum length of around 16" to 18". A radial drill press has about the same length capacity as a bench-top drill press, but it has a much larger throat capacity—which means it can handle wider pieces.

Drills & Drill Presses

NON-IMPACT DRILLS

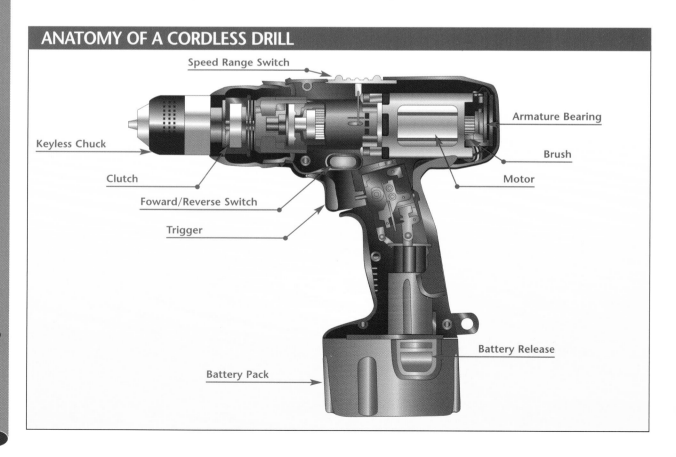

Cordless drills

There's no doubt about it, using a cordless drill is habit-forming. No more getting tangled in extension cords, or losing power because you stretched the cord and pulled out the plug. There are also safety advantages: No power cords to accidentally fall in a puddle of water, no cords catch on protrusions when you're working on a ladder and cause you to lose your balance, and so forth. So why not just go cordless? Like most things, a cordless drill is not without disadvantages.

First, you have to recharge the battery constantly. If you have only one, you'll have to stop and wait for it to recharge. If you have more than one battery, you'll be constantly swapping batteries. This might not sound like a big deal, but it can be when you're up on a roof screwing down sheeting and your battery dies.

The second disadvantage is more serious. With a corded drill, you have unlimited power (as long as you stay plugged in). On the other hand, the battery on a cordless drill is continuously being drained during use. So performance degrades as the battery loses power. Depending on the application, this can be a real hassle.

ANATOMY OF A CORDLESS DRILL

Speed Range Switch

Keyless Chuck

Clutch

Foward/Reverse Switch

Trigger

Armature Bearing

Brush

Motor

Battery Release

Battery Pack

Corded drills

OK, if a cordless drill is so convenient, then why would I even consider a corded version? The answer is power, in terms of both duration and magnitude. On a corded drill, you have unlimited power; there are no batteries to swap. Swapping batteries is a particular nuisance for jobs where sustained power or high torque is needed, such as mixing up drywall mud or driving in deck screws.

Because they're so powerful and capable of such high torque, most larger capacity corded drills come with an optional side handle to give you a more positive grip—and the added leverage you'll need to handle them. For the ultimate positive grip, select a corded drill with a D-handle or spade handle.

Another feature that corded drills offer over their cordless cousins is a power lock. A power lock allows you to "lock" the trigger in the ON position. This is useful when mixing paint, grinding, sanding, or polishing.

One final advantage of a corded drill is weight. Since a corded drill doesn't have battery packs, it can weigh less than a cordless drill rated with the same power and capacity—a big factor if all family members are planning on using the drill.

ANATOMY OF A CORDED DRILL

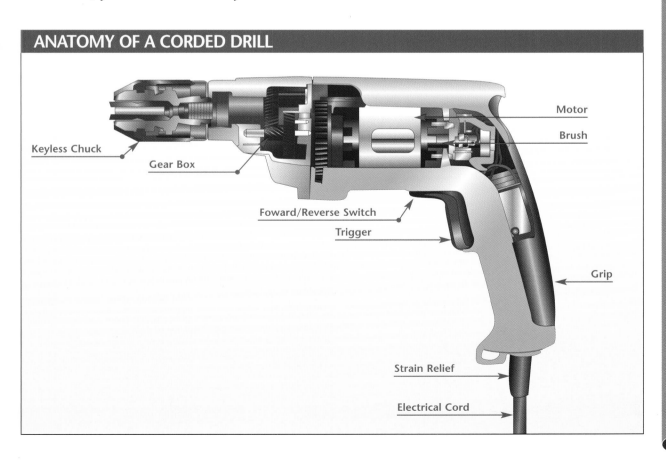

Keyless Chuck

Gear Box

Motor

Brush

Foward/Reverse Switch

Trigger

Grip

Strain Relief

Electrical Cord

FEATURES

Whether you're looking to purchase an impact or a non-impact drill, there are several features common to both drill types that can affect your purchase decision: the type of power (corded or cordless), the motor or battery ratings (and charging system), available speeds and ranges, chuck capacity and type, and whether the drill has a clutch or not.

Choosing among corded drills is simple because they don't offer as many features as the cordless variety. No bells and whistles, just raw power. Your toughest choices here are motor size and chuck capacity. If you're planning on using a corded drill pretty hard, buy one from a reputable manufacturer that offers replacement parts.

In addition to the features listed here for cordless drills, keep an eye out for extras thrown in by manufacturers to sweeten the deal. They'll often include an extra battery (worth around $40), a carrying case, or even bits. These extras may help make an otherwise difficult decision easy.

Batteries One of the best indicators of a cordless drill's power capabilities is the size of its battery. Virtually all cordless products use rechargeable Ni-cad (nickel-cadmium) batteries based on 2.4-volt "cells"—a 7.2-volt battery is made up of three cells, and so on. The more cells in a battery, the higher the voltage and generally the more powerful the drill.

Battery sizes vary from a low of 7.2 volts up to a whopping 24 volts. Likewise, higher-voltage batteries usually provide greater torque. There are two main styles of batteries: short, squat rectangular types that attach to the base of the handle, and long, thinner types that are inserted into the handle itself.

Charging systems The type of charging system a cordless drill uses will also affect its performance. Charging times vary from overnight to 15 minutes, but most systems recharge in a couple hours.

Unless you absolutely need to recharge a battery over a coffee break, stay away from the 15-minute chargers. One of the problems with these is that they overheat the battery by charging it so quickly. Heat like this will shorten the battery's life (often by as much as one-half).

Motor ratings Although toolmakers use various descriptors to indicate the "power" of a drill (heavy-duty, industrial, etc.), the true indicator of a tool's power is its amperage rating. As a rule of thumb, the higher the amperage (noted on the side of the drill), the more powerful the drill is.

The other thing to look for is the maximum torque (or twisting force) that the drill can deliver. You can usually find the drill's torque rating in the owner's manual. Higher torque is especially desirable when driving screws or drilling into tough materials.

Variable speed There's another aspect of "power" to both cordless and corded drills that muddies the water: drill speeds and ranges. In most cases, the lower the speed, the higher the torque. Speed ranges for drills vary usually from 0 to 500 rpm up to 0 to 2,500 rpm.

Many drills offer two speed ranges so that you can select between high-torque/low-speed and low-torque/high-speed to match your specific application. A cordless drill with both ranges is usually marketed as a "driver/drill."

DO I NEED A CLUTCH?

One of the most hyped and often misunderstood features of a cordless drill is the absence or presence of a clutch. A nice feature if you're planning on driving screws, a clutch is designed to help drive the screws flush with a surface. Basically, when the drill is in "drive" mode and the desired torque is achieved, the clutch slips and the bit stops turning. For convenience, most drills have a collar near the chuck that can be rotated to change settings.

Check to make sure that the collar spins freely yet still locks positively into the different settings.

In keeping with the "more is better" philosophy, tool manufacturers now offer drills with up to 25 clutch settings. In my experience, a dozen settings is plenty; any more and I end up spending too much time spinning the clutch trying to find the "perfect" setting.

Chuck capacity Standard drills are most commonly classified by their chuck capacity, or the maximum diameter bit the chuck can accept. For example, a ½" drill will take any bit up to and including ½". Shown *from left to right:* a ¼" drill, a ⅜" drill, and a ½" drill.

When portable drills first appeared, a ¼" chuck was standard. Nowadays the minimum you'll find is a ⅜" chuck. Hang on to that old ¼" drill: It'll probably be a collector's item someday.

Although many larger bits can be purchased with reduced shanks, the beefier the shank of a bit is, the better it will be able to stand up to tough jobs. I recommend that you plan ahead and go with the drill that has the larger chuck capacity.

TYPES OF DRILL CHUCKS

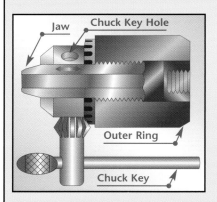

Jaw
Chuck Key Hole
Outer Ring
Chuck Key

All drill chucks work on the same principle: A set of three jaws inside the chuck housing move up and down on a set of gears to grip or release the bit (*left*). The difference is how the chuck is tightened. A keyed chuck (*center*) requires a chuck key to tighten the bit in place; a keyless chuck doesn't (*right*). There's no contest: Get a keyless chuck. You'll never miss searching around the shop for that missing key. An added benefit of a keyless chuck is that it grips the bit tighter than a keyed chuck.

Even after assessing all the various features, there's no substitute for picking up a drill to see how it fits in your hand. How is the balance? Are the controls accessible? Does it feel right? These are personal ergonomic questions that only you can answer. I've seen plenty of folks change their mind about a drill by just picking it up.

Non-impact drills are configured in one of two ways: pistol-grip or T-handle (*see below*). Virtually all corded and impact drills are pistol-grip. Cordless drills, on the other hand, can be either.

Next, take a look at the switch locations. There are four switches on a drill that can alternately make the drill a joy or a chore to use: the forward/reverse switch; the speed range switch; if it's cordless, the clutch setting collar; or, if it's corded, the power-lock switch.

Another ergonomic factor is weight. The heavier a drill is, the more fatiguing it is to use. Unless you're built like Popeye, you'll probably want to stay away from the high-voltage cordless drills (14.4-volt and above). These drills typically weigh around 5 pounds. Also, if friends or family members ever need to use the drill, a lighter weight can be used by a wider range of people.

T-handle The T-handle design evolved as batteries for cordless drills got larger and heavier; coupling a large, heavy battery with a pistol-grip made the drill tip off-center, which made it difficult to hold upright.

So tool manufacturers positioned the weight closer to center, creating a "T." This balanced the drill so well that the T-handle has become the preferred configuration in cordless drills.

Pistol-grip Although you may find T-handle drills better balanced, the handle in the center will prevent you from exerting direct force in line with the bit (unless you push the back of the drill housing with your other hand).

Since many drilling tasks need to be done with one hand, a pistol-grip drill (when held near the top of the housing) will let you get your weight behind the drill and exert plenty of force (such as when driving deck screws). Because of this, a pistol-grip drill is often referred to as an in-line drill.

Side handle To help with overall balance and provide a more positive grip, most larger-capacity corded drills are available with an optional side handle.

If you're looking to buy one of these larger drills, look for one that at a minimum allows you to position the handle on either side of the drill. Even better, look for one where the side handle can be rotated and locked into any position. This way you'll be able to adjust it for the best control—a real knuckle saver in tight quarters, like under a sink or inside a cabinet.

Forward/reverse switch The most common type of forward/reverse switch you'll find in a portable drill is a plastic arm located directly above the trigger (*see the top drill in the photo at right*); the arm toggles back and forth to change the drill's direction. As long as the arm is sturdy and its action is smooth, this is an excellent system.

Another type is a push-button system, again located above or near the trigger (*see the bottom drill in the photo at right*). If the forward/reverse switch is located anywhere other than near the trigger, look for another drill—it'll just be too awkward to use.

Speed settings Although this feature is not used as often as the forward/reverse switch, it's worth looking at the location of the speed-range switch and how easy it is to use. There are two things to look for: a large, accessible button that can be easily changed with one hand, and one with a positive lock. The drill shown on the left has a large speed switch that's much easier to toggle back and forth than the one on the right.

Note: Some corded drills have a variable-speed knob that's part of the trigger. I don't recommend these, because it's just too easy to hit the knob accidentally and change speeds.

Power lock (for corded drills only) If you're leaning toward buying a corded drill, take a close look at the power lock. A power lock allows you to lock the drill "on" for sustained use. Look for a switch that's accessible and easy to engage.

Safety Note for left-handers: Watch out for power locks on the left side of the drill housing, like the one shown in the photo. In normal use, your left hand will cover and depress the lock. When you release the trigger, the drill won't stop—this is an accident waiting to happen.

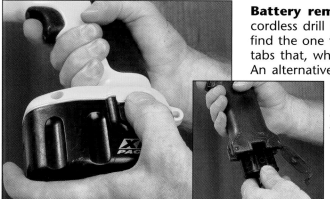

Battery removal It's worth the effort when shopping for a cordless drill to pop the battery in and out of several drills to find the one that feels best to you. Most batteries have plastic tabs that, when squeezed, release the battery (*see the photo*). An alternative system uses a metal clip to hold the battery in place (*inset*). For someone with smaller hands, the metal clip may be a better choice, since some of the larger batteries can be downright difficult to remove.

RECOMMENDATIONS

We all hate to hear it, but "How much are you willing to spend?" is a question you'll need to answer when buying a drill. The old saying "you'll never regret buying quality" certainly applies here. What you need to be careful of is overbuying—buying more tool than you'll need (or be comfortable with).

The common buying philosophy of "the more power, the better" doesn't always work. I have a good friend who scurried out to the local home center to buy the newest, most powerful drill on the market. Unfortunately, it collects dust on his tool shelf because it's so heavy and bulky that he can hardly lift it (but boy, it sure is powerful).

For general housework, I recommend a 9.6-volt or 12-volt cordless drill with a keyless chuck, two batteries, and a well-constructed case. Then, for those jobs that require a more powerful drill, buy a ½" or ¾" hammer drill with hammer-drill and drill-only modes and with two speed ranges. Between these two drills, you can tackle almost any job.

IMPACT DRILLS

An impact drill combines percussion with a rotary action so that it can drill holes in tough or abrasive materials like brick, cement, and concrete. There are two types: a hammer drill, and a rotary hammer drill; *see page 17.* Which one is best for you depends on where you live and on what size holes you intend to bore. I'm sure it sounds strange, but depending on where you live, the concrete can range from soft to very hard; *see the map on page 52.*

If you live in an area with tough concrete, or if you're planning on drilling holes that are greater than ¾" in diameter, you'll need a rotary hammer drill. On the other hand, if the concrete is soft in your area, or if you're drilling holes less than ¾", a hammer drill will do just fine.

Hammer drills are classified by their drilling capacity: the largest hole they can drill in concrete. For average around-the-house use, a ½" hammer drill is sufficient. If you're planning on a major remodeling job, select one of the larger, ¾" hammer drills.

For holes up to 1½" diameter, a rotary hammer drill with an SDS chuck will do the job. If you need to drill holes larger than 1½", the drill will probably come with a spline chuck. You may want to rent one of these, as they're quite expensive—$750 to $1,000.

ANATOMY OF AN IMPACT DRILL

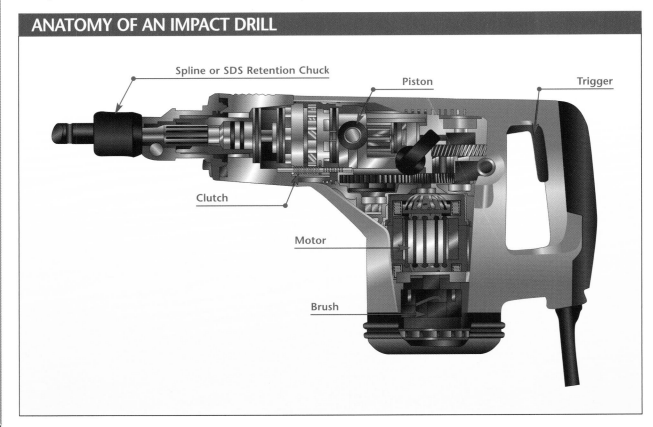

Spline or SDS Retention Chuck

Piston

Trigger

Clutch

Motor

Brush

Modes/ranges To get the most out of a hammer drill, make sure it has at least two modes: hammer/drill and drill-only mode. (Rotary hammers often have a hammer-only mode that's useful for demolition work.) Typically there are small icons near the switch that indicate drill-only (a small drill) or hammer-mode (a small hammer).

Look for a drill with at least two speed ranges and percussion ranges. The most common ranges are 0 to 1,000 rpm coupled with 0 to 20,000 bpm (beats per minute), and 0 to 2,500 rpm coupled with 0 to 50,000 bpm.

Depth stop Since you'll most likely want to use a depth stop when drilling into masonry, make sure the drill has a stop built in and that it's made of metal—avoid plastic depth stops, as they just don't stand up to prolonged use.

Also, make sure the adjustment mechanism is accessible and easy to use; I prefer a large thumbscrew-type knob like the one shown in the photo. Another feature to look for is a depth stop that can be used on either side of the drill.

HAMMER DRILL VS. ROTARY HAMMER DRILL

The big difference between a hammer drill and a rotary hammer drill is the way blows are delivered to the drill bit. In a hammer drill (*right*), the blows are the result of rubbing together two ridged metal plates (much like rubbing two poker chips together). Although it doesn't sound like much of a blow, when this occurs at high speeds (up to 50,000 bpm, or blows per minute) and it's coupled with a rotary motion, you can drill quite effectively into masonry. A rotary hammer drill (*left*), on the other hand, uses pneumatics (a piston in a chamber) to strike the bit with a cushion of air. Although these blows occur less frequently on a rotary hammer than on a hammer drill (usually less than 1,000 bpm), they're so much harder, the bit can quickly bore through even the toughest materials (and with much less vibration than with a hammer drill).

CHOOSING A PORTABLE DRILL

Although there are hundreds of drills available, they can all be classified as either an impact drill or a non-impact drill. A non-impact drill, which has only rotary action, can handle materials that are not so tough, like wood, nonhardened metals, plastics, and softer masonry (such as cinder block). An impact drill combines percussion with a rotary action so that it can drill holes in tough or abrasive materials like brick, cement, and concrete.

What's best for you: impact or non-impact? The answer lies in how often you plan on drilling into masonry. If you intend to do a lot of this, an impact drill is the best choice: It lets you bore into masonry using a fraction of the time and effort that you would with a non-impact drill.

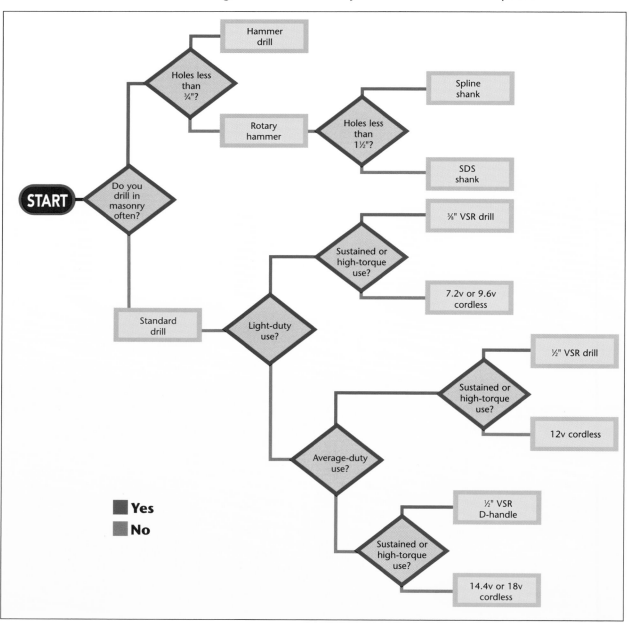

A floor-model drill press offers the largest work capacity of all the drill-press types. You can safely drill into a workpiece as long as 48". Although it's not likely you'll need this capacity often, you'll be glad you have it when you do need it. (The work capacity of a drill press is the distance from the drill chuck to the base.)

Most drill presses have an adjustable table that can swing out of the way so you can rest a long workpiece directly on the base. This table slides up and down on a hollow-steel column.

Better-quality drill presses have a rack-and-pinion gear system to move the table accurately; this type of gear system also holds the table in place when the table clamp is loosened. Without a gear system, you need to hold the table when you loosen the table clamp—otherwise gravity will take over and the table will quickly slide down the column, possibly causing damage as the table quickly drops to the floor.

The head of the drill press, which contains the motor, quill, and chuck, is mounted to the top of the column; the base mounts to the bottom.

Bits are gripped in a three-jaw drill chuck that spins inside a movable "quill." Handles on the side of the head allow you to move the quill up and down.

The amount of quill movement is referred to as the stroke of the drill press and indicates how deep you can drill into a workpiece in a single pass.

Inside the drill head, you'll typically find a set of stepped pulleys and one or two V-belts. By moving the V-belt(s) to various positions on the pulley(s), you can select different speeds. Some drill press manufacturers make a version with a continuously variable speed feature, but these can be quite costly.

ANATOMY OF A FLOOR-MODEL DRILL PRESS

- V-Belt/Pulley Cover
- Power Switch
- Quill
- Chuck
- Table
- Motor
- Feed Handle
- Rack-and-Pinion System
- Table Height Adjustment Handle
- Column
- Base

BENCH-TOP DRILL PRESS

Don't be misled by the term "bench-top" when it's used to describe a drill press. Most of us envision a light-duty, scaled-down version of a large stationary tool. Although some bench-top drill presses are just that, many aren't. As a matter of fact, you can find some bench-top drill presses that are much more powerful and have more desirable features than a full-sized floor-model drill press.

The major difference between a bench-top drill press and a floor-model drill press is the work capacity—the distance between the base and the drill chuck. This is typically 16" on a bench-top drill press and is 44" to 48" on a floor model. Other than that, a quality bench-top drill press can do just as much as a floor model.

With that in mind, you might ask, "Why bother with a bench-top? Why not get a floor model?" The answer has to do with floor space and storage options. A floor-model drill press takes up a certain amount of floor space, known as its "footprint." But a bench-top drill press also has a footprint and takes up floor space if it rests on anything but a workbench (where it will also take up valuable space). Since most of us aren't willing to give up any bench space, we usually buy or make a stand for the bench-top version.

The advantage to this is that although you'll still lose floor space, you'll gain storage space—as long as the stand is more than a set of legs. Floor space is something that a floor-model drill press doesn't offer. And we all know what a premium storage space is in a workshop.

Quick Tip: One of the least expensive stands you can find is a "scratch and dent" three- or four-drawer bathroom or kitchen cabinet. Check your local home center to see if they have a clearance section of "as is" merchandise. Sometimes you can pick one of these up for just a few dollars. Then all you'll need to do is add a plywood top.

ANATOMY OF A BENCH-TOP DRILL PRESS

- V-Belt/Pulley Cover
- Power Switch
- Motor
- Quill
- Chuck
- Feed Handle
- Table
- Table Height Adjustment Clamp
- Column
- Base

Although a radial drill press has many of the same parts as a bench-top or floor-model drill press, a single glance will tell you it's a different breed of cat. A radial drill press is about the same size as the bench-top version, but with one huge difference: It has a much larger throat capacity (typically around 36").

That's because the head of the drill press is attached to a horizontal ram column that can slide in and out. This column is similar in size and function to the main column, which supports the head. On some models the ram column moves in and out via a rack-and-pinion mechanism.

For angled drilling, the head and ram column can be rotated either to the left or to the right. Most models also allow you to pivot the head on the main column so that you can drill compound-angle holes easily.

Sounds like a great idea, yes? So why aren't radial drill presses more popular? A couple of reasons. First, with all the extra parts required and the additional manufacturing time, radial drill presses are more expensive than the other type—you can expect to pay around 50 percent more for a radial drill press.

But more importantly, radial drill presses have a reputation for not being as accurate as a standard drill press. A common complaint is that the bit tends to go out of square when throat capacity is adjusted.

The other thing that I've always felt was odd about this style of drill press is that although the ram column allows you to pull the head out from the main column, the table is still small. This means that you'd need to build a rock-solid table extension in order to drill accurately with the head extended out so far.

ANATOMY OF A RADIAL DRILL PRESS

- V-Belt/Pulley Cover
- Power Switch
- Quill
- Chuck
- Feed Handle
- Table
- Column
- Motor
- Ram Column Lock Lever
- Table Height Adjustment Handle
- Rack-and-Pinion System
- Base

FEATURES

Just like selecting a drill, choosing the right drill press is a matter of identifying the type of work you're planning on doing and then comparing the available features of different models. For most of us, the feature that comes to mind immediately is the capacity of the drill press, in terms of both chuck to column and throat capacity (*see below*).

But there are other features that are just as important. You'll also want to look closely at a couple of things. First, the table: its type, size, and ability to tilt. Second, the speeds available and the method of changing them. Most drill presses use an induction motor, and for good reason: They run at a constant speed. Although it may seem inconvenient to have to move a belt to change speeds, you can be assured that the speed you choose will be constant.

This is not true of a drill press that uses a universal motor. If the drill press you're interested in varies the speed electronically, then it has a universal motor. These motors tend to drift significantly off speed while in use, especially under load. Stick with an induction motor and pulleys—you'll have fewer headaches in the long run.

Size (chuck to column) Drill presses are sized according to their throat capacity—the distance from the center of the chuck to the column. This is a case where bigger is better.

The confusing thing about this is that drill-press manufacturers double this measurement when they advertise their products. For example, a drill press with a throat capacity of 6" is marketed as a "12" drill press" because it can drill to the center of a 12"-wide board.

Table The type, size, and features of the table on a drill press can have a huge impact on its usability. The main types are round, and square or rectangular. Go with the square or rectangular type, as it's easier to clamp a workpiece to it. As to size, pick the largest table you can afford, since it'll offer the most support for a workpiece. Finally, make sure that the table tilts for drilling angled holes and that it is easy to adjust.

The drawing at right illustrates: (1) throat capacity, (2) stroke, (3) length capacity, and (4) table size.

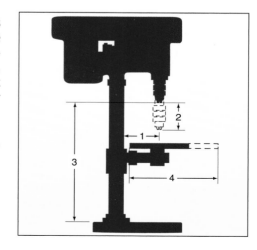

Speeds Almost every drill press is capable of running at different speeds, typically between 20 and 3,600 rpm (revolutions per minute). Most use an induction motor that runs at a constant speed; to change speeds, you have to move a V-belt on stepped pulleys.

Many bench-top models offer 5 speeds, while most floor models come with at least 12; *see the chart below.* The larger the selection, the easier it is to match the speed to the material you're drilling into. Look for large, easy-to-turn thumbscrews or levers that loosen the motor mount for changing speeds.

Belts/pulleys Flip the cover off the drill press head and examine the belts and pulleys used to change speeds. You're looking for a couple of things here.

First, check to make sure the pulleys are well made and the casting are smooth and burr-free. Then turn one pulley and watch the other; any wobble is a sign of impending vibration problems. Finally, go through the procedure for changing speeds: It needs to be quick and easy to do.

COMMON DRILL-PRESS SPEEDS

Speed	RPM Settings
5-speed	620; 1,100; 1,720; 2,340; 3,100
6-speed	425; 600; 1,100; 2,050; 3,900; 5,500
8-speed	375; 600; 900; 1,275; 1,800; 2,400; 3,200; 4,200
9-speed	150; 260; 300; 440; 490; 540; 1,150; 1,560; 2,200
12-speed	250; 360; 410; 540; 590; 650; 1,090; 1,280; 1,450; 1,820; 2,180; 3,000

Table adjustment: Slip collar The tables on all drill presses move up and down to adjust the position of the workpiece in relation to the drill chuck. Many older models and some inexpensive modern models use a slip collar, like the one shown here.

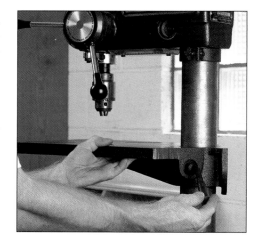

Basically, when a bolt or thumbscrew is loosened, the table is slid up or down on the column. The disadvantage to this is that the table is heavy and has to be supported when it's moved.

Table adjustment: Rack and pinion A big improvement over a slip-collar table adjustment is an interlocking rack-and-pinion system that supports the table at all times. Adjustment can be made with one hand: Just loosen the locking mechanism and turn the crank on the table to move it smoothly up or down.

Note: This type of table adjustment also makes it easy to fine-tune the position of the table—something that's quite difficult to do with a slip collar.

Table tilt You can drill angled holes on the drill press by tilting the table. (If your table doesn't tilt, check out the shop-made tilting table on *page 88*.) It's worth the time to check out what you have to do to get it to tilt.

On some models, you simply loosen a knob or handle, tilt the table, and tighten the knob. Other versions, like the one shown here, require you to loosen and remove a stop before the table can be tilted.

Once you've identified the features you want, all you have to do is find a drill press that has them. Simple, right? The problem is that you'll find many drill presses with similar features. This is where ergonomics comes into play.

The best way to pick between similar tools is to run each through its motions, like a test drive. Go through each procedure: Adjust the speed, lock the quill, set the depth stop, raise and lower the table— the differences will likely be apparent.

As a rule of thumb, adjustments that use a hinged lever instead of a knob are better. A lever lets you exert more force as you loosen or tighten a locking mechanism.

The need for extra leverage has to do with vibration. Even the highest-quality drill press vibrates. This can cause the locking device to self-tighten so that it's hard to loosen. This is especially true with a knob—it just doesn't provide the mechanical advantage that a lever does.

Table lock Since you'll be constantly changing the position of the drill press table to compensate for different-sized work-pieces, the location and style of table lock will have a large impact on day-to-day work.

The first thing to look for in a table lock is accessibility. You should be able to reach it easily with either hand. Second, the best leverage is provided by the lever style, as shown. Stay away from knobs, as they just don't offer sufficient purchase.

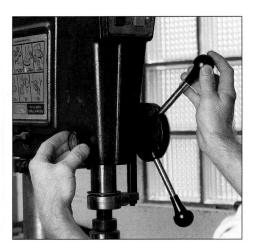

Quill lock A quill lock allows you to freeze the quill in any position along the length of its stroke. This is particularly useful when using a drum sander. Most quill locks pinch the quill to secure it in place.

There are two main types of quill locks. One style is incorporated into the depth-adjustment ring. I prefer the other style (*as shown*), which has a separate lever-type handle. All it takes is one quick turn of the handle to lock the quill.

Depth stop A well-designed depth stop adds precision to your drilling. The two main types are a rotating depth scale and stop nuts on a threaded rod.

Rotating depth scales are common on newer drill presses and are quicker to adjust than the threaded rod type (where you have to spin the nuts up and down the rod to change settings). Here again, look for a lever-style lock on the rotating type instead of a knob.

Feed handles The quill-feed handles should be comfortable to use and easy to remove. Look for a drill press with a minimum of three feed handles that have soft rubber balls on the ends.

Make sure the feed handles thread into the feed shaft hub. Unlike a feed handle that's held in place with a cotter pin, threaded feed handles can be quickly removed for the occasional drilling job where they interfere with the workpiece. Note: If you find that your feed handles tend to loosen with use, you can prevent this by wrapping a couple turns of Teflon tape around the threads and then screwing them back in.

RECOMMENDATIONS

For all-around general use, I recommend a bench-top drill press with a minimum of five speeds and a ¾-hp induction motor. It'll handle almost any job, and you can use the space below it for storage.

If you're a woodworker, look for a drill press with a slow speed (around 250 rpm)—many specialty bits (like Forstner bits) work best at lower speeds. Make sure that the speeds are easy to change. If you know you'll be working on long or tall stock, a floor-model drill press will be a better choice.

When you've narrowed down your choices to just a few, keep a couple things in mind. First, stick with a name brand you can trust. Reputable tool manufacturers that have been around for decades are still in business for a reason: They make quality products. And, just as important, you can get replacement parts, even for their older machines. Second, a "bargain" tool usually isn't a bargain. It's cheaper because it was manufactured with looser tolerances and fewer steps. Take a look at the castings and the machined surfaces—they'll be clean and smooth on a quality drill press.

CHOOSING A DRILL PRESS

In my opinion the big choice you'll have to make when choosing a drill press is what's more important, capacity or storage space. If capacity is foremost, a floor-model or radial drill press is your best bet. If you're planning on drilling into tall or long workpieces, go with a floor model; if you'll need to drill to the center of wide workpiece, a radial drill press is the logical choice.

Bench-top drill presses offer many of the features of the floor-models, but they sacrifice capacity (in terms of chuck to base) to provide you with the option of storage space. If your shop is like mine, storage space may be more valuable than the extra capacity a floor model offers. If this is the case, a bench-top drill press will serve you well.

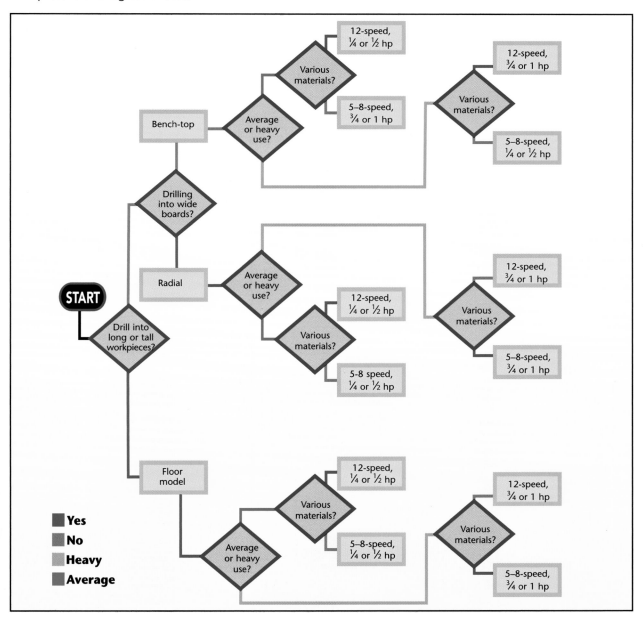

Drills & Drill Presses

27

2 ACCESSORIES

Even the most powerful, state-of-the-art drill or drill press can't get the job done without the correct accessory. In this chapter, I'll show you ways you can boost the performance of your drill or drill press by selecting the right accessory for the job, from common bits to drill stands.

Naturally, the accessory you'll reach for most often when using a drill or drill press is a bit. The five most common types are: twist bit, spade bit, brad-point bit, Forstner bit, and masonry bit. Each one excels at a specific task and has its own advantages and disadvantages. (*See pages 29–33 for more on this.*)

But there's a whole lot more out there than just bits. To fine-tune your drill's performance, consider specialty bits like countersink bits, pilot bits, self-centering bits, and hole saws.

Then there are sanding and grinding—two of the more everyday tasks that a drill or drill press is called on to perform. Accessories to handle these jobs include drum sanders, disk sanders, flap sanders, grinding wheels, and grinding points.

Safety Note: You should always wear eye protection when you use your drill. I can't emphasize enough how important this is when you're sanding and grinding; both of these operations spew out dangerous amounts of dust and debris.

Other tasks a drill or drill press are well suited for are polishing and buffing. Here again, numerous accessories, including wire wheels, brushes, buffing wheels, and pads, can make these chores more manageable. **Safety Note:** Just as with sanding, grinding, and all drill operations, please protect your eyes.

Next, I'll take you through accessories that were created to add precision to your portable drilling jobs: stands, guides, and dowel jigs (*pages 38–39*). Finally, there are numerous accessories you can buy to enhance the performance of your drill press: plug cutters, circle cutters, rotary surface planers, and mortising attachments (*pages 40–41*).

The real challenge comes when you're at the local hardware store or home center—which one do you buy? To help with this daunting task, I'll share some buying tips as I review each accessory so that you'll be able to tell the good from the bad and the ugly.

Twist bits Twist bits can drill into just about anything because they're ground to a "universal" angle, usually 118° or 135°. When buying bits, look for high-speed steel and a reputable manufacturer. Stay away from bargain bits; they may be inexpensive, but they won't last. Some manufacturers increase bit longevity with coatings. The most common is titanium-nitride (identifiable by a bright gold color), which increases resistance to wear and abrasion.

There are three designation systems used for twist bits (*see the chart below*): numeric, from #80 to #1 (.0135" to .228"), alphabetic, from A to Z (.234" to .413"), and fractional, from ¹⁄₆₄" to ¹⁄₂", in ¹⁄₆₄" increments. The numeric (often referred to as wire gauge) and the letter system are bits that fall between standard fractional sizes.

Brad-point bits Brad-point bits excel at drilling round, clean holes in wood. On a quality brad-point bit, the spurs protrude from the lips. This is the way a true brad-point bit is designed to work: The spurs score the wood, and the lips pare it away.

On the bargain bits, the spurs and lips are ground in the same plane; they cut at the same time and there's no pre-scoring. The result is more tear-out and a ragged hole.

LETTER/NUMBER DESIGNATORS FOR TWIST BITS

Drill	Dia.	Drill	Dia.	Drill	Dia.	Drill	Dia.	Drill	Dia.	Drill	Dia.	Drill	Dia.	Drill	Dia.
80	.0135	64	.0360	48	.0760	33	.1130	18	.1695	3	.2130	9/32	.2812	W	.3860
79	.0145	63	.0370	5/64	.0781	32	.1160	11/64	.1719	7/32	.2188	L	.2900	25/64	.3906
1/64	.0156	62	.0380	47	.0785	31	.1200	17	.1730	2	.2210	M	.2950	X	.3970
78	.0160	61	.0390	46	.0810	1/8	.1250	16	.1770	1	.2280	19/64	.2969	Y	.4040
77	.0180	60	.0400	45	.0820	30	.1285	15	.1800	A	.2340	N	.3020	13/32	.4062
76	.0200	59	.0410	44	.0860	29	.1360	14	.1820	15/64	.2344	5/16	.3125	Z	.4130
75	.0210	58	.0420	43	.0890	28	.1405	13	.1850	B	.2380	O	.3160	27/64	.4219
74	.0225	57	.0430	42	.0935	9/64	.1406	3/16	.1875	C	.2420	P	.3230	7/16	.4375
73	.0240	56	.0465	3/32	.0938	27	.1440	12	.1890	D	.2460	21/64	.3281	29/64	.4531
72	.0250	3/64	.0469	41	.0960	26	.1470	11	.1910	E	.2500	Q	.3320	15/32	.4688
71	.0260	55	.0520	40	.0980	25	.1495	10	.1935	1/4	.2500	R	.3390	31/64	.4844
70	.0280	54	.0550	39	.0995	24	.1520	9	.1960	F	.2570	11/32	.3438	1/2	.5000
69	.0292	53	.0595	38	.1015	23	.1540	8	.1990	G	.2610	S	.3480		
68	.0310	1/16	.0625	37	.1040	5/32	.1562	7	.2010	17/64	.2656	T	.3580		
1/32	.0312	52	.0635	36	.1065	22	.1570	13/64	.2031	H	.2660	23/64	.3594		
67	.0320	51	.0670	7/64	.1094	21	.1590	6	.2040	I	.2720	U	.3680		
66	.0330	50	.0700	35	.1100	20	.1610	5	.2055	J	.2770	3/8	.3750		
65	.0350	49	.0730	34	.1110	19	.1660	4	.2090	K	.2810	V	.3770		

Spade bits Although used primarily in construction trades for drilling rough holes in wood, a spade bit (often referred to as a paddle bit because of its shape) is just the ticket for many around-the-house jobs. How can you tell a good spade bit from a bad one? Look for spurs. Older spade bits, like the one shown, have none.

Newer spade bits have a set of cutting spurs on the outer corners of the blade. These score the perimeter of the hole before the center is scraped away, resulting is a much cleaner hole.

Forstner bits Forstner bits can handle problematic drilling jobs like drilling clean holes at a steep angle, partial holes, and even overlapping holes. Instead of being guided by a centerpoint like other bits, a Forstner bit is guided by its rim. The rim scores the wood while a pair of lifters plane away the waste. The result is a clean, flat-bottomed hole. When shopping for these bits, look for a combination of high-speed steel and a small centerpoint.

The original bits desgined by Benjamin Forstner were designed for use in a hand brace at slow speeds. Even today, these bits require low speeds to keep from overheating the delicate rims.

HOLE PROFILES

In woodworking and carpentry, it's often necessary to drill a "stopped" hole—one that doesn't go all the way through a workpiece, typically when a piece of hardware needs to be recessed below the surface of the workpiece. Some bits are better for this task than others. Shown, *from left to right,* are a spade bit, brad-point bit, twist bit, and Forstner bit.

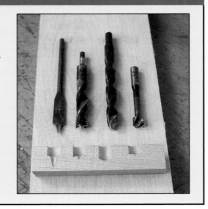

The ultimate bit for the job is a Forstner bit, as it leaves a flat-bottomed hole and has a small centerpoint; the flatter the hole, the better it can support a washer and distribute weight more evenly when a bolt or screw is tightened. Brad-point and twist bits are the second best choices—they too have small centerpoints and produce a relatively flat bottom. A spade bit, with its large centerpoint, should be used only when the workpiece is thick or when it will break through the other side.

Multi-spur bits A popular derivation of the Forstner bit is the multi-spur bit. Unlike a Forstner bit, which uses a rim to score the perimeter of the cut, a multi-spur bit uses a set of jagged teeth. The gullets between the teeth efficiently whisk away chips. This means the bits can be run at higher speeds and will stand up to heat better.

But because multi-spur bits don't have a continuous rim, they don't cut angled, partial, or overlapping holes anywhere as cleanly as a Forstner bit can. Unfortunatley, many mail-order woodworking catalogs market these bits as Forstner bits.

STORAGE OPTIONS: INDEX AND SLEEVES

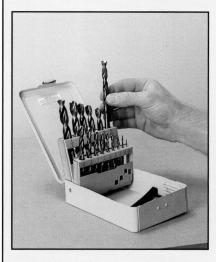

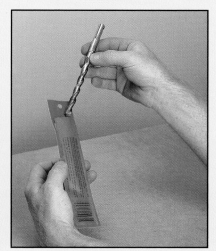

One of the worst things you can do to a drill bit is to toss it into a drawer or box with other bits where they can roll around and ding the flutes and dull the cutting edges. If you don't have an index like the one shown *above* for your drill bits, get one. They're relatively inexpensive and are worth the investment.

Individual plastic sleeves like those shown *above* also do a good job of protecting your bits. Indexes and sleeves will not only prolong the life of the bits, but also keep them well organized. The only catch here is that you've got to use them—try to get in the habit of returning your bits after every use.

MASONRY BITS

Carbide-tipped The business end of a masonry bit is carbide-tipped to drill holes in abrasive materials like brick, cinder block, concrete, and tile. The type of bit you use will depend on the drill and the type of chuck it has.

For a standard drill or a hammer drill with a three-jaw chuck, you can use any masonry bit that has a round shank. If it's a hammer drill or a rotary hammer drill, the chuck may take an SDS or spline shank bit; *see below.*

Shanks styles: round, SDS, spline Round shank masonry bits that fit in three-jaw chucks afford the worst grip since there just isn't anything for the jaws to lock onto (*top bit in drawing*). This is especially true when a round shank bit is used in a hammer drill: The constant percussion can often vibrate the chuck jaws loose. I've seen this type of bit actually fall out of a drill chuck in use.

Two better gripping systems are SDS (*center bit in drawing*) and spline (*bottom bit in drawing*). Although a tool clerk may tell you that SDS stands for "special-direct-system," it originally stood for "Steck-Dreh-Sitzt." Roughly translated from German this means "insert–turn–locked-in." Indentations in the shank provide the chuck with a firm purchase to guarantee that the bit won't come flying out in use.

With the spline shank system, lengthwise slots offer much more surface area for the transmission of power, while at the same time affording a rock-solid grip. You'll only usually find this type of shank style on the larger heavy-duty or industrial-strength roatry hammer drills.

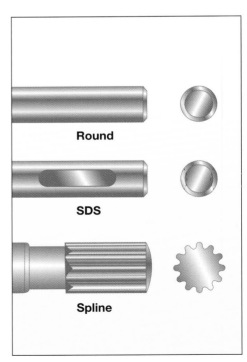

Round

SDS

Spline

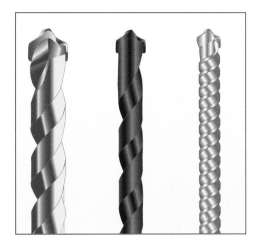

Flute patterns A noticeable difference between masonry bits is their flute patterns; in particular, their spacing. As a general rule of thumb, the closer together they are, the faster material is removed from the hole.

But there's more to it than this; the larger the hole you're drilling, the larger the debris that the flutes have to carry out. Closely spaced flutes work fine on bits under ½", but choose a bit with wider spacing for larger holes, like the left one shown in the drawing.

Tip configurations The tips on most masonry bits are shaped like a twist bit (*left*); other manufacturers use a protruding tip to help prevent the bit from "walking" as the hole is started (*right*).

When examining bits, check to make sure the carbide is brazed cleanly to the bit. Stay away from bits that show signs of pitting or an uneven, rough weld. Don't skimp on quality here. As a general rule of thumb, the thicker the carbide, the longer the bit will last.

HANGING ON TO MASONRY BITS

The bits used in hammer drills and rotary hammer drills really take a beating. Not only are the materials they drill into tough, but also the drills themselves are literally pounding on the bits as they rotate. You'd think this constant percussion would eventually loosen the chuck and create a dangerous situation. It can. That's why many drill manufacturers have developed special chucks (and bits) to eliminate this possibility (*see the opposite page*). *Shop Tip:* If you're using an SDS or spline bit, the gripping system will last much longer if you apply a little white lithium grease to the end of the shank before inserting it in the chuck.

SPECIALTY BITS

Countersink Typically used after a shank or pilot hole has been drilled, a countersink bit creates a depression in the material to fit the head of the screw.

There are two types: The most common (*at right in photo*) has a series of flutes to scrape the material and form the depression (often producing a scalloped cut); the other type (*at left in photo*) has a single cutting edge that slices the wood and leaves a much cleaner hole.

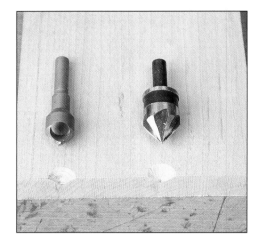

Pilot bits A pilot bit (*at far right in photo*) makes installing screws painless because it drills the pilot hole, shank hole, countersink, and (if desired) a counterbore all at the same time. A special kind of pilot bit, the screw pilot bit (*at center left and center right in photo*) allows you to adjust the length of the pilot hole but not the length of the shank hole.

Taper pilot bits (*at left in photo*) offer a separate countersink and a stop collar that slide up and down on a bit tapered to match the profile of a standard wood screw. These typically adjust with an Allen wrench (*inset*).

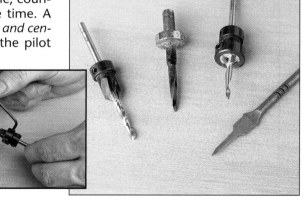

VIX BITS: INSTALLING HINGES WITH EASE

A self-centering (or vix) bit is a totally reliable way to install a hinge without the usual skewing and misalignment. The "magic" of this bit is an inner and outer sleeve that spin around a twist bit. When the tip of the self-centering bit is inserted in a hinge hole and depressed, an inner sleeve retracts up into the outer sleeve (*see drawing*). This positions the twist bit so it can drill a perfectly centered hole for the hinge screw.

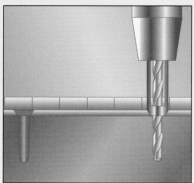

Hole saws When you need to drill holes greater than 1½" in diameter, reach for a hole saw. The most common type (*at right in photo*) consists of three parts: a two-piece arbor, consisting of a shank that accepts a twist bit to guide the cut, plus a nut that threads on the end to hold an interchangeable metal cup in place. The set of metal cups, usually made of carbon steel or bimetal, have teeth ground or stamped into the edge to do the cutting.

Heavier-duty hole saws, like the one shown *left* in the photo, have only two pieces: The arbor is welded onto the cup; the bit can be replaced in case it breaks. Regardless of the type, it's essential that you use slow speeds and stop often to blow out sawdust that can build up and cause burning.

Hole-saw variations Carbon steel hole saws are intended for use in wood, plastic, and composite materials that are less than ¾" thick. The different diameter saws use the same interchangeable arbor set (*the top hole saw in drawing*).

Designed for industrial use, bimetal hole saws can drill into stainless steel and mild steel, cast iron, copper, brass, aluminum, hardwood, plastics, and composites (*the center hole saw in drawing*). As you might expect, bimetal hole saws are expensive; you can buy a set of carbon steel hole saws for the price of a single bimetal saw. But if used properly, these will far outlast the carbon steel variety.

Plug-ejecting hole saws have a built-in spring that automatically ejects the plug after the hole has been cut and eliminates the hassle of prying the plug out (*the bottom hole saw in drawing*).

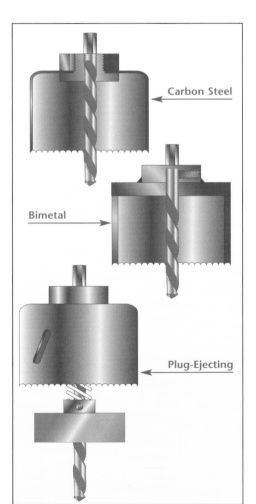

Carbon Steel

Bimetal

Plug-Ejecting

Drum sanders A drum sander is a rubber cylinder with an arbor running through it that when tightened expands to grip a sanding sleeve. Drum sander sizes vary from ½"×½" up to 3"×3" and can be used either in a drill press (*photo*) or with a portable drill (*inset*). Sanding sleeves are available in three basic grits: coarse, medium, and fine.

When shopping for a drum sander or a kit, pay extra for the type with longer bodies. Not only do these allow you to sand thicker materials, but also the larger sanding surface helps your sanding sleeves last longer since you can distribute wear and tear over the entire length of the sanding sleeve.

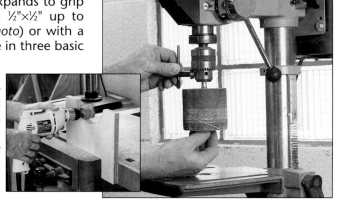

Disk sander A disk sander is just a rubber disk (with an arbor) that accepts a sanding disk. The disk can either have a hole in it for attaching to the disk via a screw and washer or it can be solid and backed with pressure-sensitive adhesive (PSA) sandpaper. Attaching the PSA-style one is simply a matter of peeling off the backing and sticking it onto the disk.

All disks are not the same. A thin rubber disk (*at upper right in photo*) is very flexible; this is great if you're sanding a curved surface. But if you're working on a flat surface, you'll be better off with a rigid disk (*at lower right in photo*). Another thing to look for is a dimpled surface—you'll find it's easier to remove PSA sandpaper from it than from one that's smooth and shiny.

Flap sander Flap sanders (sometimes referred to as flapwheels) are great for sanding hard-to-reach convex and concave surfaces, slight profiles, and rounded areas. Most flap sanders come mounted on a ¼" or ⅜" spindle.

Each wheel is made up of hundreds of abrasive cloth sheets attached to a core. As the wheel spins, fresh sanding grit is constantly being exposed to the workpiece—you can remove a lot of wood in a very short time with one of these. This can be both good and bad. I recommend keeping used flap sanders around for finer work; fresh grit can be a bit aggressive.

Wire wheels and brushes A wire wheel fitted with an arbor or a wire brush can quickly remove rust and old paint from metal. Wire wheels are usually 5" in diameter and usually come in two "grits": fine and coarse.

Wire brushes also come in fine and coarse and in a couple of different diameters (typically, 1¾" and 2¾") and are usually fitted with ¼" shanks. These accessories are inexpensive and disposable, so just buy a brand name that you trust.

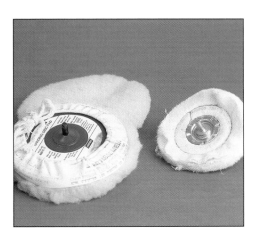

Buffing wheels Buffing wheels combined with a little jeweler's rouge will rapidly bring a high luster to metal tools, jewelry, and similar items. Usually made of muslin, buffing wheels come in 3", 4", and 6" diameters.

Buffing pads (or bonnets) are designed to slip over a rubber disk. You'll find them in diameters ranging from 5" to 10". The pad itself may be made from lamb's wool, or more commonly a mixture of wool and polyester. Both materials work well.

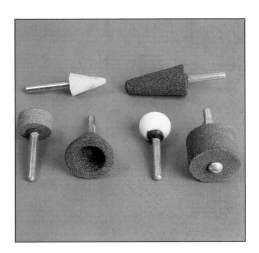

Grinding wheels and points For all-around use, a sharpening wheel fitted with an arbor works well. Typically 5" in diameter, these wheels are available in fine, medium, and coarse (100, 60, and 36 grit, respectively).

Grinding points like those shown in the photo, on the other hand, usually only come in one grit, but in an amazing array of shapes. Particularly well suited for detail work, most grinding points are fitted with a ¼" shank for use in either a die grinder or an electric drill. Smaller, more delicate points come with a ⅛" shank.

JUST FOR PORTABLE DRILLS

Vertical stand A vertical stand like the one shown in the photo holds a portable drill in a carriage to basically convert your drill into a drill press. There are a couple things to look for when shopping for one of these.

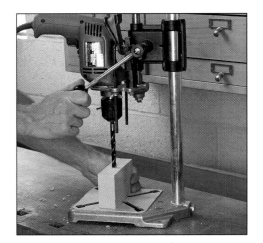

First, make sure that the stand can hold your drill (ask the tool merchant) and that it's easy to install and remove the drill. (Note: Vertical stands are designed to hold pistol-grip drills only.) Second, to get any kind of precision, the stand must firmly and positively lock the drill in place. Look for a clamping mechanism that adjusts easily so you can square up the bit.

Adjustable drill guide An adjustable drill guide (often referred to as a Portalign, which is a brand name) turns a portable drill into a precision drilling machine. No more question as to whether a hole is perfectly 45°, 55°, or 90° anymore. With this guide, it is.

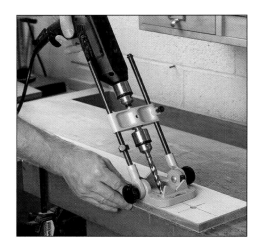

The large base of the guide provides a stable platform and easily adjusts between 45° and 90°. (For a shop-made version of a sliding drill guide, *see page 84.*)

Quick-change bits I remember the first time I saw quick-change bits. I thought to myself, who's in that much of a hurry that they can't take the time to loosen a chuck and change a bit? Then I thought about assembling a project when I'm constantly changing between a drill bit and a drive bit, and I started to see the light. It's not that I'm in that much of a hurry, it's just a lot more convenient.

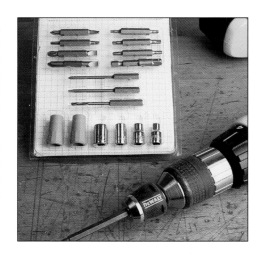

Most quick-change bit systems are based on a special arbor, like the one shown in the bottom of the photo, that allows you to remove a bit simply by pulling down on the outer collar. Insert a new bit and release—it's that easy. The only drawback with most systems is that they accept only hex-shank bits.

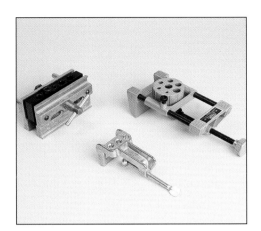

Dowel jigs Dowel jigs clamp to a workpiece and support the drill bit to ensure the you drill perfectly straight holes in both pieces to be joined with a dowel. Make sure the jig you buy has the hole sizes you need; inexpensive versions only have a few—typically ⅜", ⁵⁄₁₆", and ¼" (*the center jig in photo*).

More expensive but higher-quality dowel jigs like the right and left jigs shown in the photo can handle ⁷⁄₁₆" and ½" holes as well. My favorite dowel jig (*at left in photo*) takes all the guesswork out of alignment: It automatically centers the hole on the thickness of the workpiece as it's tightened.

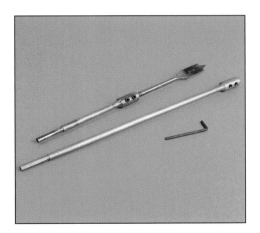

Extensions A drill extension is a long metal rod with a holder at one end to accept spade bits with ¼" hex shanks. Two Allen-head setscrews lock the bit in place.

Extensions are available in various lengths (6", 12", and 18"). Not only do they provide access to awkward locations, but they're also great for drilling deep holes that are ⅝" or over in diameter (otherwise there won't be sufficient clearance for the bit holder).

Right-angle drive For drilling holes in awkward or tight places, nothing beats a right-angle drive. As the name implies, a right-angle drive is an attachment for your portable drill that lets you drill at 90°.

In addition to providing access to tight spaces, this accessory also improves handling when you're sanding or buffing (especially if the drive comes with an optional handle, as shown here).

JUST FOR THE DRILL PRESS

There are a number of drill accessories that should only ever be used on the drill press; in particular: plug cutters, circle cutters, rotary surface planers, and mortising attachments. Each of these accessories requires a number of things that only a drill press can offer.

First, all of these accessories need to be perfectly perpendicular to the workpiece—an easy task for a drill press but virtually impossible with a hand-held drill. Any deviation from perpendicular can result in a serious accident: A cutter or rotating blade can dig in and grab the workpiece (and possibly your hand if it's close by).

Second, every accessory discussed here requires slow, steady speeds to operate efficiently. Again, easy for a drill press, much tougher for a portable drill. Finally, most also need a powerful motor to drive them through the toughest of hardwoods. Not a problem for the ¾-hp or larger motor on a drill press.

Plug cutters You can make your own wooden plugs with a plug cutter to cover the counterbored screws in your woodworking and home improvement projects. The advantage to using a plug cutter instead of buying manufactured plugs is that you can cut the plugs from the same wood as your project and match up the color and grain.

Plug cutters can cut either straight plugs or slightly tapered plugs. I prefer tapered plug cutters because once it's installed, the plug creates a virtually invisible patch.

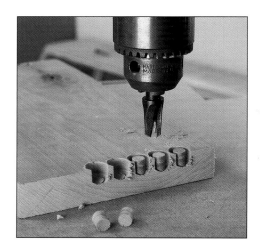

Circle cutters It's easy to cut large holes (up to 8") with a circle cutter and your drill press. A circle cutter is made up of an arbor and bit and one or two adjustable cutters (*left and right in photo, respectively*).

In use the arbor, bit, and cutter(s) must be absolutely perpendicular to the workpiece. Because of this, circle cutters should never be used in a portable drill. Portable drills also can't provide the power and slow, steady speed that these large bits need to remove the copious amount of wood that the cutters remove as the hole is cut.

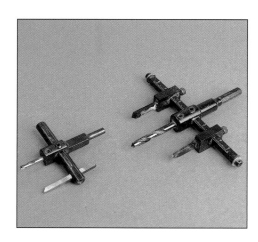

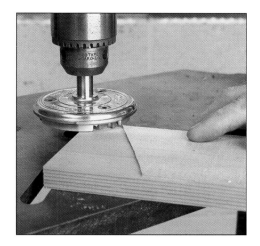

Rotary surface planer If you ever need to plane a piece of wood that is too thin or too short to safely put into a planer, a rotary surface planer for your drill press is just the answer.

The one shown here, sold under the brand name Safe-T-Planer, has a set of three cutters inset into the head that can safely remove up to 1/64" of wood in a single pass. It's also handy for cutting rabbets, bevels, tapers, and decorative cuts.

In use, the planer is chucked into the drill press and the table is raised to determine the depth of cut. Keep your hands clear of the head, and feed the workpiece into the planer against the direction of rotation.

MORTISING ATTACHMENT AND BITS

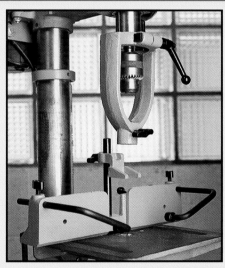

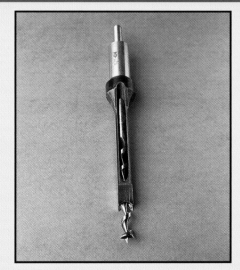

A mortising attachment for the drill press lets you cut square holes or "mortises" for joining together wood parts. How does it "drill" square holes? Basically it has to do with the special bits. The thing to look for in a mortising attachment is well-machined parts that are easy to install and set up on the drill press. (*See pages 74–75* for more on this.) Most attachments come with a hold-down bracket that presses the workpiece against the table and holds it steady as the bit is raised.

The secret to "drilling" square holes with a mortising attachment is the bits. An auger-type bit rotates within a square-edged hollow chisel.

The bit cuts a round hole, and the chisel punches the corners square. Both parts must be very sharp for this to work. Chisels come in a variety of sizes; the most common are 1/4", 5/16", 3/8", 1/2", and 3/4".

3 BASIC DRILLING OPERATIONS

All of the basic operations that you can perform with a portable drill or a drill press (regardless of the type) can be broken down into a series of simple tasks. The most fundamental of these tasks is installing and removing bits. Although this seems pretty obvious, there are a couple of tips that can help ensure that your drill bits get installed quickly and properly so that you'll never encounter a "flying" bit (*see page 43*).

Over the years, I've noticed that many woodworkers and homeowners alike think that drilling success will be guaranteed as long as they choose the right bit for the job. Not so. Drilling requires more than matching a bit to a task or material—you have to know what the bit needs in terms of speed and feed rates to cut properly.

In some cases, you can use the same bit for a variety of materials; twist bits can handle wood, metal, and plastic (*pages 48, 50, and 54, respectively*). But different materials require different drill speeds, feed rates (how slow or fast the bit enters the material), and lubricants. Other materials, like masonry, glass, and tile, require special bits and conditions to drill holes efficiently (*pages 52 and 55*).

Knowing what a bit needs to drill effectively will help to ensure successful drilling, but it won't guarantee precision or accuracy. No matter what type of drill you're using, accuracy begins with measuring, layout, and marking (*see pages 44–45*). Granted, not all jobs require pinpoint accuracy; but for when you encounter a job that does, I've included a number of simple tricks for precision. For portable drills, see *pages 58–59;* for the drill press, see *pages 60–62.*

Finally, I'll show you a couple of ways to drill holes with a portable drill wherever you want without creating a big mess.

Keyed A drill with a keyed chuck uses a small L- or T-shaped key to tighten or loosen the jaws: clockwise to tighten, counterclockwise to loosen. **Safety Note**: Don't depress the drill's trigger while holding the outer ring of the drill chuck to tighten the chuck around the bit; the last thing you want is a flying bit.

Also, if your drill has an electrical cord, get in the habit of unplugging it whenever you change a bit. One of the best ways to force yourself to do this is to attach the chuck key to the electrical cord near the plug so that you'll have to unplug the drill to use the key.

Keyless A keyless chuck grips a bit without the necessity of a key. Not only is this convenient (you won't waste any more time searching for that ever-evasive chuck key), but it's been my experience that this type of chuck actually grips a bit tighter than a keyed chuck.

To use a keyless chuck, hold the base of the chuck with one hand and spin the outer body of the chuck with the other until the jaws grip the bit. Then give the base and the outer body of the chuck a final sharp twist in opposite directions to snug the jaws tight around the bit.

Drill press Although you can find keyless chucks for a drill press, most come standard with a keyed chuck. To install or remove a bit, insert the chuck key into one of the holes in the drill chuck, engage the teeth, and rotate the chuck key clockwise to tighten, couterclockwise to loosen.

For the chuck jaws to get a positive grip on the bit, most manufacturers recommend inserting a bit at least ½" into the chuck. To get the firmest grip possible, slip the chuck key in each of the holes in the chuck and tighten. This ensures that all jaws come in full contact with the shank of the bit.

Drills & Drill Presses

43

PREPARING TO DRILL

The steps you take before you drill a hole can have as much impact on the results as the choice of drill and bit and the way they're used. Drilling a hole exactly where you want depends first and foremost on how well you measured and laid it out. If you take the time to accurately measure and mark the location, you've won half the battle. The other half has to do with giving the bit a "leg up" to start the hole where you marked; this is most often in the form of a centerpunch (*see below*).

Finally, you'll rarely experience any kind of precision or accuracy if the workpiece itself isn't held firmly in place. With a portable drill, small pieces should get clamped in a vise; larger pieces can get clamped directly to a workbench or sawhorse.

If you're working on a drill press, this means always taking the time to clamp your workpiece to the drill press table or holding small pieces in a vise or other clamp.

1 Measuring and layout Wood and masonry can be marked with a pencil or a felt-tip marker. For other materials such as plastic or metal, it's more accurate to scribe the layout marks with a sharp awl. Use a try square (*as shown*) or combination square to transfer measurements accurately.

For best accuracy when drawing lines from a mark, don't position a try square or combination square on the mark. Instead, try the more accurate trick that draftsmen have been using for centuries: Place your pencil or awl on the mark and gently slide the square up against the pencil. Now, holding the square firmly in place, draw the line.

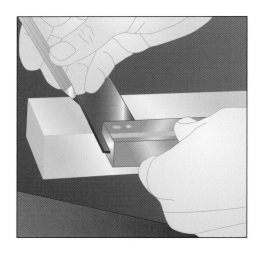

2 Centerpunch Centerpunching creates an indentation in the material to guide the centerpoint of a bit. A centerpunch will work in most materials with a light blow from a mallet or hammer.

A word of caution here. If you're drilling into brittle materials such as tile or slate, it's best to make a dimple in the material instead of centerpunching it—you'd run the risk of cracking and ruining the piece. Dimpling is easily done with a drill and a cone-shaped grinding point, or even a sharp countersink.

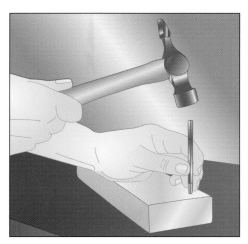

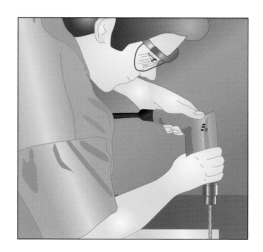

3 **Protective eyewear** After you've laid out and center-punched the hole location in the material, it's almost time to drill. But first, a word of warning: I can't stress enough the importance of using protective eyewear whenever you drill.

Although I'm sure it looks harmless, even the smallest drill bit is capable of slinging wood shavings, metal filings, or bits of concrete into your eyes. Please wear your safety glasses when drilling!

4 **Secure workpiece for portable drill** It's also imperative that the workpiece you're drilling into is secure; that is, that it can't break free and rotate if the bit catches.

If you're working on a small piece, hold it with a small vise or a handscrew. For larger workpieces (like the one shown here), clamp the workpiece securely to a workbench, sawhorse, or table with C-clamps, pipe clamps, or whatever you have on hand.

5 **Clamp workpiece to drill press table** When using a drill press, many folks simply hold a workpiece on the table with their hands when they go to drill into it. But even if you've got an iron grip, a spinning bit can catch a workpiece and rip it out of your hand. At the very least, it'll slap your knuckles. At worst, it's a projectile.

Take the few seconds necessary to firmly clamp your workpiece in place with stout clamps, especially when drilling into small parts and metal.

CHOOSING THE RIGHT SPEED

Drill speeds

The type of material that you're drilling into will affect the drill speed, the feed rate (amount of pressure you apply), and the type of lubricant (if any) you use. As a general rule of thumb, the harder the material, the lower the drill speed. (*See the drill speed chart below.*)

In terms of drill bit size, the smaller the bit, the greater the speed needed for effective cutting. Conversely, as the bit gets larger, the drill's speed needs to decrease.

But there's more to drilling a hole than selecting the correct speed. The feed rate (how much pressure you apply to the drill) is also important. What may not be obvious is that the feed rate directly affects the bit's cutting speed.

Let's say, for instance, you're drilling into a piece of softwood with a ¼" twist bit at a recommended speed of 1,000 rpm. If you were to drill five holes varying the feed rate from 1" to 5" per second, you'd get cleaner holes at the lower feed rates. Why? Because although the bit speed is constant, you are changing how quickly the bit cuts into the wood. In effect, you're increasing the bit speed away from its recommended speed.

Although this can get confusing, there's a really simple way to determine both correct speed and feed rate: Just listen to the motor as you drill. The sound that the motor makes is your best indication of proper combination of speed and feed rate.

Start with the recommended speed for the material for the size hole you're drilling. As you begin to drill, listen to the drill's motor. If it starts to bog down, back off on the feed rate. If it sounds like it's free-running and not really cutting into the material, increase the feed rate.

It's just like listening to your lawnmower as you mow the lawn. If you go too fast, it bogs down; too slow, and you're not cutting effectively. With a little practice, you'll quickly learn the "right" sound that your drill makes when it's drilling at the correct combination of speed and feed rate.

Finding the correct speed and feed rate for a cordless drill can be more of a challenge than for a corded drill because most cordless drills have two speed ranges: a low speed range for high-torque applications such as driving screws, and a high speed range for most drilling operations.

Unless you're using a large bit or drilling a deep hole, use the high speed, low-torque setting. This will make it easier in the long run to listen to the drill's motor and "learn" the correct speed/feed rate for a given task.

SPEEDS & MATERIALS

Material	Speed (in rpm)
Softwoods	800–2,400
Hardwoods	600–1,800
Ferrous metals	100–900
Nonferrous metals	300–2,400
Masonry	100–800
Glass	300–2,400
Plastics	300–1,400

1 **Changing drill-press speed: Loosen belt tension** To change the speed on a drill press, you'll first need to locate the belt-tensioning adjustment or lock-down if there is one (consult your owner's manual).

Loosen the knob or setscrew, and with one hand on the motor and the other on the front of the drill head, squeeze your hands together to remove the tension from the belt or belts.

2 **Changing drill-press speed: Move belt** If you don't have a belt-tensioning mechanism, you can change speeds by first slipping the belt off one step of the pulleys. Turn one pulley by hand to help the belt ride up or down the steps. If you do have a belt tensioner, and you loosened it in Step 1, the belts can easily be moved from one pulley to the other.

When you've moved the belt or belts to the desired position, reapply tension (*see page 105* for more on this), and tighten the knob or setscrew to lock the belts in place.

BELT AND PULLEY POSITIONS

Changing speeds on most drill presses is done by moving a belt or belts to various positions on a set of "stepped" pulleys. The slowest speed is obtained by placing a belt on the largest step on the motor pulley and the smallest step on the quill pulley. Conversely, the highest speed occurs when the belt is on the smallest step on the motor pulley and the largest step on the quill pulley. You'll usually find a chart inside the pulley cover illustrating the various pulley positions and their respective speeds.

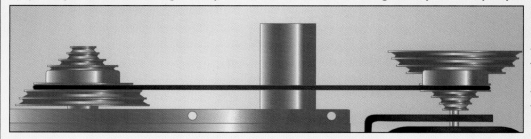

DRILLING IN WOOD

For the most part, drilling into wood is simplicity itself. No lubricant is necessary, and the speed and feed rate will be determined by the size and type of the drill bit and the type (and species) of wood you're drilling into. This can be hardwood, softwood, plywood, or any of the other manmade composite products.

The biggest challenge occurs when you're drilling into softwood or hardwood and you hit a knot. You'll know immediately: The drill will bog down and often catch. When this happens, reverse the drill and back out of the hole. More than likely the bit will be clogged, particularly if you're drilling into a softwood like pine.

Once you've cleaned the bit with a brass brush, slip it back into the hole and continue drilling at a slightly higher speed but at a slower feed rate. This lighter cut creates finer chips, which are less likely to clog. Continue drilling and cleaning the bit periodically until the hole is drilled.

Shavings are a great indicator In addition to the sound the drill motor makes, there's another quick indication that you've hit the right combination of drill speed and feed rate: the wood shavings themselves. If they're uniform in size and the bit isn't clogging, you've selected an appropriate speed/feed rate combination.

If you're not getting nice curly shavings (like those shown at left in the photo) and you're sure your bit is sharp, continue searching for the right speed/feed rate combination.

Backer board for a portable drill One challenge to drilling wood is preventing tear-out. Tear-out occurs when a drill bit breaks through the opposite side and the fibers of the wood are not supported (*inset*).

A dependable way to prevent this is to back up the workpiece with a backer board. Simply clamp a scrap piece of wood to your workpiece and drill through your workpiece and into the scrap. Since the workpiece is supported by the scrap, no tear-out or splitting occurs.

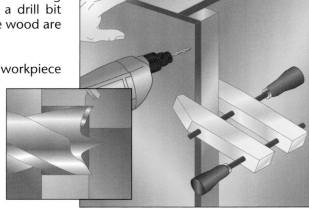

Backer board on drill press You should also use a backer board whenever you're drilling through a workpiece on the drill press. Simply slip a scrap of wood under the workpiece before drilling.

The only thing to be careful of here is to make sure the center hole in the metal drill press table is centered roughly underneath your bit. This way if you accidentally drill through the backer board, you won't ruin the bit by hitting a metal table.

DRILLING IN BOTH DIRECTIONS

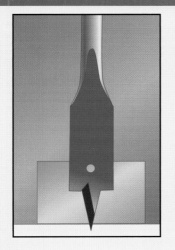
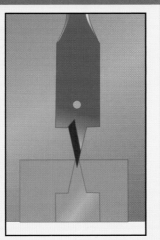

Another way to prevent tear-out is to drill into the workpiece from both directions. This is a particularly useful tip when using drill bits with a long centerpoint (such as a spade bit shown). To do this, drill into the workpiece until the point just begins to break through the opposite side. Then use this indent to start your drill bit from the opposite direction. If any tear-out does occur, it happens inside the hole and is unseen.

DRILLING IN METAL

Drilling in metal can be as simple as drilling in wood, as long as you follow a few basic rules. First, make sure that you use a sharp bit. You can get away with drilling into wood with a not-so-sharp bit, but not metal—it just doesn't work. Second, it's imperative that you centerpunch to prevent the bit from wandering. Third, use slow speeds and the proper lubricant.

As a general rule of thumb, the softer the material, the higher the speed to cut effectively. If you're drilling a ½" hole in aluminum, your speed should be around 2,300 rpm; for hard steel, 500 rpm would be appropriate.

The type of lubricant you use will depend on the material you're drilling into (*see the chart below*). The primary purpose of the lubricant is to keep the bit cool by reducing heat-generating friction, which can damage both your bits and your workpiece.

A lubricant also helps to keep the hole clean by floating away small metal particles and shavings that can clog up the spirals of a twist drill. Don't get carried away with lubricant, though. A small puddle the size of a penny will do in most cases; any more and you'll end up with a shirt or shop apron that looks like you've been hanging around the spin art booth at the local fair.

Corkscrew shavings Just like drilling into wood, you'll know you're at the correct speed and feed rate for metal when the shavings peel off in long, tightly coiled corkscrews.

Safety Note: Metal shavings of any kind are extremely sharp. Always use a brush to clean off your drill bits and workpiece. In addition to preventing a nasty cut, this also lessens the likelihood of picking up one of those annoying metal slivers that's so tiny you can't even see it—but you sure can feel it (sound familiar?).

LUBRICANTS	
Material	**Lubricant**
Wood	None
Hard steel	Turpentine
Machine steel	Soluble oil
Wrought iron	Lard oil
Brass	Kerosene
Glass	Kerosene
Aluminum	Kerosene
Plastic	None

Securing the workpiece There are a couple of really good reasons that you should always securely clamp a metal workpiece before drilling. First, drilling creates friction and the resulting heat can quickly make the workpiece too hot to handle. Second, twist drills have tendency to "catch" the material as they exit and spin the workpiece—even throw it.

Clamping a workpiece securely to a drill press table can be difficult because the underside of the table isn't flat: The castings often get in the way. One nifty solution to this problem is the locking pliers shown here. The base of the pliers attaches to any of the slots in the drill press table. Once in position, this can quickly secure a workpiece.

A Lubricant well Lubricants can't do their job if they don't stay where they're needed—at the cutting edge of the bit. To keep lubricants (especially thin ones) where they can do some good, make a small well or "moat" around the bit.

Plumber's putty works great for this. Roll some into a coil and wrap it around the bit, leaving about ½" between the putty and the bit. Drizzle in a small amount of lubricant, and drill.

DRILLING INTO HARDENED METALS

Most twist bits can't drill into metal that's been hardened or "heat-tempered." Carbide bits can, but they're expensive and prone to breakage. Here's a trick that'll let you use regular bits. Chuck a piece of metal rod in the drill. Then turn it on and press the spinning rod down onto the workpiece where you want to drill (*left*). Continue until the metal heats up and discolors—what you're doing here is reversing the heat treatment and removing the temper. Now you can easily drill into the softer metal (*right*).

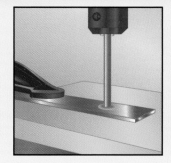

DRILLING IN MASONRY

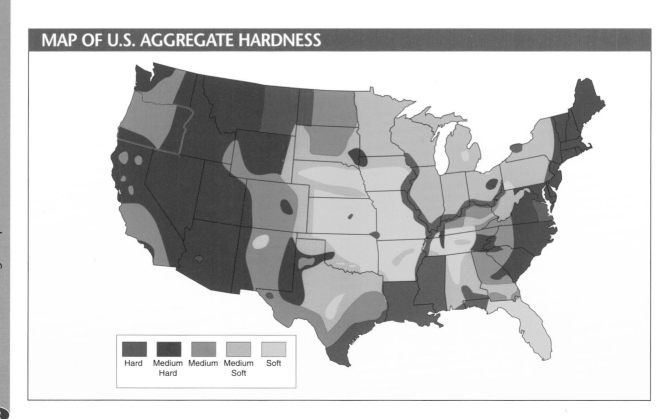

If you're just drilling a couple of small holes in soft masonry (like cinder block), you can get by with a non-impact drill and a carbide-tipped masonry bit. But if you've got a lot of holes to drill, if they're larger than ¼", or if the material is hard (solid brick or concrete), you'll need a hammer drill or a rotary hammer drill.

As discussed in Chapter 1, the criteria for deciding which type of drill to use are what size the hole is and whether you're drilling into concrete. Concrete will range from soft to very hard, depending on where you live; *see the map below.* (The variations in hardness come from the different local aggregate—loose rock—

they mix in with cement to form concrete.)

To drill into masonry using a hammer drill, first flip the drill into hammer-drill mode. Then insert an appropriate drill bit: round, SDS, or spline-shank (*see page 32*).

Once you start drilling, let the tool do the work. Applying excessive force won't make it drill any better; you'll just get tired a lot faster. You do, however, need to apply sufficient pressure to

prevent the tool from bouncing, or "dancing" as they call it in the trade. This really is a matter of trial and error, but you'll know that you've hit the correct pressure and feed rate when you see a smooth, even flow of dust coming out of the hole.

MAP OF U.S. AGGREGATE HARDNESS

| Hard | Medium Hard | Medium | Medium Soft | Soft |

Set depth stop Most hammer drills have a built-in depth stop to help you accurately drill to the correct depth. To use a depth stop, loosen the adjustment knob and slide the stop rod out to the desired depth. Then tighten the knob and begin to drill, easing up on the feed rate as the rod gets close to the surface.

Because percussion is involved with a hammer drill, it's important to stop when the rod hits the surface. If you don't, the vibration can damage the stop rod or the locking mechanism.

Blow out dust As with most drilling operations, it's a good idea to stop often, remove the bit, and clear out dust and debris from the hole. This also allows the bit to cool. Although compressed air works best to blow out the hole (as shown), you can also use lung power.

Don't be tempted to use water as a lubricant or to try and cool the bit. In addition to creating a slurry that clogs the flutes of the bit, the water will only serve to cushion the blows of the drill—exactly the opposite of what you want.

Notched guide When drilling with a hammer drill, you're dealing with a combination of percussion and rotary action. This can make guiding a bit straight a real challenge. Your best bet is to make a simple drilling guide by cutting a V-notch in a 2×4, as shown.

Also, if you're planning on drilling a large hole in masonry, I've found that it's best to start with a smaller bit and work your way up. Not only does this make it easier to start the bit, but it's also easier on the drill, the bit, and you.

DRILLING IN PLASTIC

Just as with wood, drilling into plastic is simple. No lubricant is necessary, and the speed and feed rate will depend on the material you're drilling into (*see the chart on page 46*). The trickiest part to drilling plastic is finding the correct speed/feed rate combination.

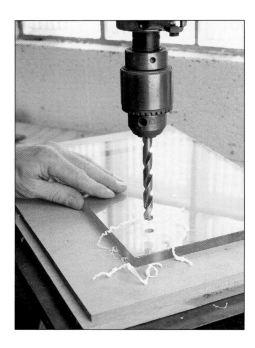

Although you won't experience tear-out in plastic as you do in wood, it is quite possible for an exiting bit to break or chip off small pieces of plastic. This is often the case when drilling into plastic laminates (such as Formica or Wilsonart). The thin layers of these are prone to chipping. To prevent this, you can use the same measures used for wood: Support the work with a scrap piece, or drill from both sides.

Because plastic will melt when sufficient heat is present, it's important to keep both the workpiece and the drill bit cool. The simplest way to do this is to frequently lift the bit out as you drill to clear out friction- and heat-generating chips. This is particularly important if you're drilling into thermosetting plastic.

If you're drilling holes to accept metal hardware (screws, nails, and so forth), it's best to drill the holes slightly oversized (around 1/64" to 1/32"). This allows for the inevitable expansion and contraction of the plastic.

Finally, if you're planning on working with plastic often, consider purchasing a set of bits designed for drilling in plastic. Although they're similar in appearance to twist bits, the tips are ground to a sharper angle, typically around 60°.

Corkscrew shavings As with metal, the shavings from plastic should peel off in tightly coiled corkscrews (*at right in photo*). If your feed rate is too fast, the plastic may chip, crack, or shatter. If the feed rate is too slow, heat may build up, resulting in a melted plastic blob with your bit stuck in it.

I'd recommend drilling a couple of test holes in a piece of similar scrap plastic before drilling into your workpiece. Also, if you need to drill a large hole, consider starting out with a smaller bit and working your way up.

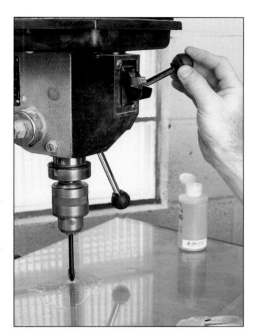

To many homeowners, the thought of drilling into glass or tile is a nerve-racking one. But it isn't that difficult. Like many other home improvement jobs, all it requires is a little know-how and the appropriate tools—in this case, specially designed glass and tile bits (*see the sidebar below*).

In addition to a special bit, there are a couple of rules for successfully drilling through glass or tile. The first rule is that the glass or tile must rest on a spotlessly clean, flat drilling surface. Even the smallest lump of dirt or a grain of sand caught under the glass or tile during drilling can cause a crack to start.

Second, the bit must be kept well lubricated; kerosene, turpentine, or even water will get the job done (although kerosene works best). Third, the feed rate or pressure you apply to the bit must be heavy and very steady to keep the bit grinding. If you see visible chips, you're not pressing hard or steady enough.

Finally, it's imperative that you stop drilling just as soon as the tip of the bit breaks through. The best thing to do then is to flip the workpiece over and continue drilling (grinding, actually) from the other side.

GLASS AND TILE BITS

Although you can drill through some tile with a masonry bit, the breakage rate is quite often unacceptably high. A better solution is to use a glass and tile bit. These bits are designed just for "drilling" through tile and glass. They use either a diamond-pointed or a tungsten carbide–pointed tip to grind their way through the glass. You can find glass and tile bits at almost any hardware store or home center, in a wide variety of sizes.

DRIVING SCREWS

Driver/drill One of the most labor-saving features of a portable electric drill is its ability to drive screws. A cordless drill with a clutch and a hefty battery—a driver/drill—is the best tool for driving screws.

A corded drill, however, may be a wiser choice for a lengthy job where you'll need a lot of torque (such as driving long deck screws into pressure-treated lumber when building a deck or gazebo).

Drill a pilot hole Regardless of the type of drill that you're using to drive screws, it's often a good idea to drill a pilot hole first. Drilling a pilot hole effectively reduces the impact of material density by removing it instead of trying to displace it with the screw.

This is basically an all-around good idea whenever you drive screws into wood, especially near the end of a board, where it's prone to splitting.

DRIVING SCREWS WITHOUT A CLUTCH

If your drill doesn't have a clutch, don't despair: There are a couple of tricks you can use to drive screws reliably. The first of these is to pulse the trigger and slowly drive the screws in small increments. I've seen a lot of folks do this even when they have a clutch. When the head of the screw gets close to the surface, finish driving the screw in at a slow speed (assuming you have variable speed). If you don't have variable speed, there are two options to finish the job. One is to just reach for a screwdriver and give the screw a final turn or two. Or, if your drill allows you to lock the chuck, you can lock it place and simply twist the drill to set the screw (it seems kind of silly, but it works).

Clutch settings If your drill has a clutch, driving screws is just a matter of finding the appropriate setting and then driving the screw flush. But getting used to clutch settings and finding the right one for the job will take some experimentation. The greater the number of settings, the better your chance of finding the perfect one. With a bit of practice, you'll be able to get within a setting or two on the first try.

Material density If you're lucky enough to be working on a project with consistent materials (like attaching drywall to metal studs), you'll be able to set the clutch and drive screws with perfect regularity.

However, when you're working with a more finicky material (such as hardwood or softwood), the variations can cause you to spin the clutch settings like a frantic teenager who has forgotten the combination to a school locker. You're best bet here is to pick an average setting that leaves the screws a bit proud, then pulse the trigger to set the screw flush. This way you won't bury the screw in the material or cause it to strip.

SCREW TYPES: TAPERED, STRAIGHT-SHANK, SQUARE-DRIVE

Tapered: Tapered wood screws (*left*), long the fastener of choice for woodworkers, require a pilot hole that's tapered or of two different diameters.

Straight-shank: Straight-shank screws (*center*) are becoming increasing popular because they require only a single diameter pilot hole.

Square-drive: A version of the straight-shank screw, square-drive screws (*right*) feature a square recess in the head, which offers a more positive drive system.

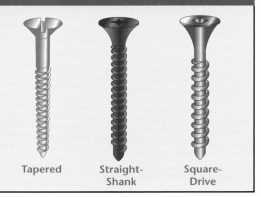

Tapered   Straight-Shank   Square-Drive

PORTABLE-DRILL PRECISION

Portable drills aren't known for their precision. Granted, there are plenty of jobs where you don't need to be super-precise—you're drilling into drywall to install plastic anchors, or drilling pilot holes in decking for screws, for example. But there are those times when accuracy counts. Sure, the size of the hole you're drilling will be accurate—choosing the correct size drill bit should take care of that. But drilling perfectly straight holes can be a challenge.

Fortunately, no matter what type of material you're drilling into, there are a number of ways you can increase the accuracy and precision of your portable drill. To drill to a specific depth, you can buy commercially made stops, or you use a simple shop-made version (*see below*). If you need to make sure the hole you're drilling is straight, there are a number of sight guides, drilling guides, and levels that you can buy or make that will make this easy.

Depth stops The most common manufactured depth stop is a metal collar that slips over the bit (*at right in photo*). A small setscrew locks the collar in place. The disadvantage to this is that tightening the setscrew can crush the flutes of the bit, resulting in poor chip ejection and clogging.

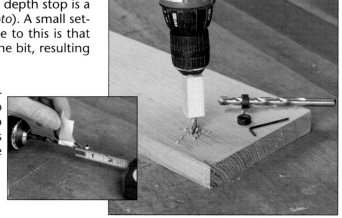

I've always gotten along with two simple shop-made versions that don't harm the bit: a scrap block of wood with a hole in it that's cut to length so just the right amount of bit protrudes (*see photo*), and a piece of masking tape wrapped around a bit (*inset*).

Sight guides Perhaps the simplest method for drilling an accurate straight hole is to use one of your layout tools as a reference. For straight holes, set a small try square or combination square alongside your drill bit and use it as a guide for your drill. Even a block of wood that's cut reliably square will work in a pinch as a simple sight reference.

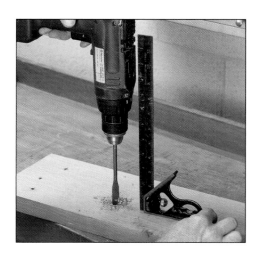

Bubble levels An alternative to a sight guide is a bubble level. Bubble levels come built-in on some portable drills and are available as "stick-on" accessories for those that don't.

These are an invaluable aid when drilling in tight or awkward quarters where it isn't feasible to use a sight guide. To use a bubble level, just position the drill so the bubble indicates that you're level, and then drill away.

Pilot holes Even if you use a sight guide or a bubble level, you're out of luck if the bit wanders off your mark. Assuming that you have marked and centerpunched your hole, it's still a common problem for a bit to wander, especially if you're using a large bit.

Fortunately, the fix is easy. Just start by drilling a small pilot hole all the way through for your larger bit to "follow."

Drilling guides The next step up in accuracy from a sight guide is a drilling guide. The concept is simple. Precision holes are drilled in a guide block to support the bit along its length. As you drill, this support prevents you from tilting the bit, ensuring the correct angle.

Just as you can purchase manufactured drilling guides, you can also use a simple shop-made guide block that will work just as effectively (*see pages 80–81*).

DRILL-PRESS PRECISION

Unlike a portable drill, where precison comes from a sight guide or drilling guide, precision is built right into a drill press. Drilling accurate holes is what it's designed for. Whether you need a perfectly straight hole or to drill to an exact depth, the drill press is the tool for the job.

But like all mechanical marvels, a drill press needs to be tuned and adjusted in order to guarantee a high level of precision. Fortunately, this only takes a couple of minutes and should be done whenever precision is paramount or anytime you adjust the table; *see below*.

There are also a number of simple tricks that you can use to make your drill press even more precise. In particular, add clamping tools, fences, and stops to your drill press table so you can quickly and accurately position a workpiece. These tricks are especially useful for repetitive work like drilling holes in identical parts.

Align the table Aligning a drill press table so that it's perfectly perpendicular to the drill chuck is a snap with this quick and easy procedure.

Chuck a 5" or 6" length of straight ¼" metal rod in the drill. Then adjust the drill press table up or down so that you can butt a small try square up against the rod. Loosen the table-tilt nut or bolt until it's friction-tight, then adjust the table until there's no gap between the blade of the try square and the rod. Tighten the table-tilt nut and check the gap again. In some cases, tightening the nut can shift the table. Repeat the procedure as necessary until there's no gap.

Check the quill If you notice that the centerpoints on your bits tend to wobble slightly when they're in the drill press, first roll them on a known, flat surface to make sure that they're not bent.

If they're straight, the quill may need some attention. You can check the quill for "runout" by using a dial indicator as shown (*see page 103* for more on this procedure).

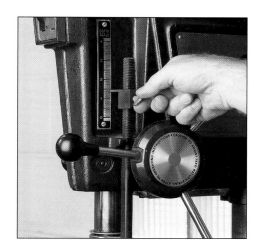

Use the depth stop Drilling a hole to an exact depth is child's play for a drill press. Position the workpiece next to the bit, and lower the quill to the desired depth marked on the workpiece.

Then adjust the depth-stop nut or the rotating depth scale to lock the quill in this position. Drill a test hole in a scrap piece and check the depth. Re-adjust as necessary.

Note: On drill presses that use a rotating depth scale, it's important to lock the rotating ring securely. If you don't, it's possible to drill too deep if you apply a lot of pressure to the feed handles as you lower the quill.

Spindle lock You can accurately lock the quill in any position along its stroke with the quill lock, usually located near the front edge of the base of the drill head. This is most often done when using a drum sander or other abrasive tool on the drill press.

It's also a precise way to adjust the depth of cut when using a rotary surface planer. *See page 41* for more on this handy accessory.

Repetitive work When you need to drill an accurate series of holes in similar-shaped parts (like a set of wheel axles for a toy train), consider this quick setup.

First, position the workpiece on the drill press table, aligning it to the bit. Then set a framing square or try square against two adjacent sides of the part and clamp the square directly to the drill press table. This forms a dual fence to position each part quickly for drilling.

Fence One of the best ways to position a workpiece accurately on a drill press is to use a fence. This can be as simple as the 2×4 shown, or as fancy as the shop-made fence on *page 86*.

In either case, place the workpiece on the table so the bit is roughly aligned with the mark on the workpiece. Then position the fence so the bit and mark are perfectly aligned, and clamp the fence in place. *See the sidebar below* for a super-accurate way to position the fence.

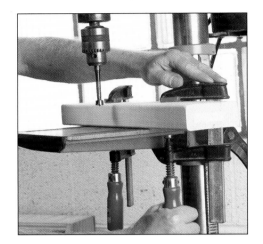

Horizontal stops Horizontal stops take the fence described above to a higher level of precision. A horizontal stop is just a scrap of wood clamped to the fence. It "stops" a workpiece from sliding from side to side when a hole is bored. Stops also let you accurately drill repetitive holes, much like the try square on *page 61*. To prevent sawdust from building up between the workpiece and stop, clamp the stop about ⅛" above the table.

POSITIONING THE FENCE WITH A DRILL BIT

Accurately positioning a fence a short distance (less than ½") away from a bit can be a real challenge. Even if you take your time and measure carefully, odds are you'll be slightly off. Here's a super-precise way to position it: Just reach for a twist bit whose diameter matches the distance you need between the fence and the bit. Then simply place the shank of the bit between the chucked bit and the fence, slide the fence over so it butts up against it, and clamp it in place. Remove the bit and you're ready to drill.

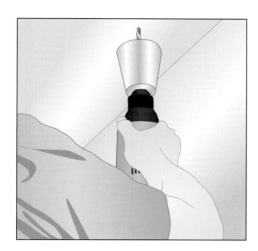

Drilling overhead No matter what you drill or where you drill it, you're going to create sawdust, shavings, or dust. Say for example you need to drill holes for a new fixture in the ceiling above your dining room table. You could take the time to spread out drop cloths, or you could use this nifty trick.

Simply poke a hole in the bottom of a paper cup or Styrofoam cup. Insert your bit and then trim the cup so that sufficient bit is exposed to drill your hole. As you drill, the cup catches the shavings or dust. Nice and neat!

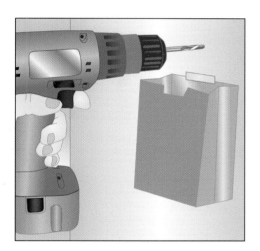

Drilling into walls As long as you're working in a shop, outside, or in a basement, you usually don't have to worry about making a mess. The floor is often concrete or tile and it sweeps up easily.

But what if you need to drill inside where there's carpeting? To avoid making a mess, just tape a paper bag to the wall under the hole to catch the shavings or dust. Just be sure to use a tape that won't damage your wall treatment, such as masking tape or painter's tape.

At the drill press Drilling holes, especially large ones, on the drill press can create a surprising amount of shavings; a sanding drum can generate a veritable cloud of sawdust.

Instead of making a mess and cleaning it up afterwards, eliminate it at the source by clamping the nozzle of a shop vacuum directly to the table on the drill press. Angle the nozzle for the optimum pickup, and secure it with a clamp or duct tape.

Drills & Drill Presses

4 ADVANCED DRILLING OPERATIONS

In addition to drilling straight holes in a variety of materials, an electric drill or drill press can handle a wide range of tasks, including drilling angled, deep, or partial holes and drilling in round stock (*pages 65, 67, 69, and 70, respectively*). The good news is that you don't need fancy accessories for the majority of these jobs—most can be handled with a little Yankee ingenuity.

Some other common tasks that portable drills and drill presses excel at are sanding, buffing, and grinding. Armed with a drum sander and a disk sander (*page 71*), you can handle almost any sanding job around the house,

from smoothing curved parts to refinishing a tabletop.

With grinding wheels and points (*page 72*), you can sharpen everything from plane irons to garden shears. Add a buffing pad to your arsenal, and you can quickly put a shine on a variety of projects—whether it's polishing a brass doorknob, buffing out the final coat of varnish on an heirloom project, or just getting a mirror finish on your classic car (*page 73*).

Holes that are larger than 1½" in diameter are manageable with a hole saw in either a portable drill or the drill press (*page 76*); holes

larger than 3" in diameter can be safely cut with ease by using a circle cutter (*page 77*) on the drill press (this is also a nifty way to make wheels for toys).

Woodworking joinery is even possible on the drill press with the addition of a mortising attachment (*pages 74–75*). A mortising attachment allows you accurately to cut a square mortise—one half of a mortise-and-tenon joint. The mortise-and-tenon joint is a time-tested way to join wood parts together with both beauty and strength.

The obvious difference between drilling a straight hole and an angled hole with a portable drill is that you need to hold the drill at a precise angle. Two easy ways to do this (*shown below*) are to use a sight aid and to use a guide block.

A not-so-obvious problem you're likely to encounter when drilling an angled hole is a wandering bit. And the more acute the angle you're drilling at, the greater the tendency the bit has to wander. Wandering like this occurs primarily with twist drills, since they offer no spur or point for starting the hole.

You can prevent a bit from wandering in a couple of ways. First, if you have brad-point bits or spade bits, use them. Both of these bit types offer an aggressive starting point that twist bits don't. If all you have on hand are twist bits, use a guide block with them (*see below*).

Sight aids As an aid to maintaining your drill bit at a precise angle, use a layout tool to sight along, just as you used a try square to drill a straight hole. But in this case you'll need a sliding bevel.

Another thing that helps is to check your progress often. Back out the drill to clear out chips, and then insert a dowel, pencil, or other thin round object in the hole (*inset*). This way you can take a step back and check the angle.

Guide blocks To prevent twist bits from wandering when drilling angled holes, you can make a guide block to support the bit as it starts the hole. Just drill a hole in a piece of scrap and cut off one end to match the desired drilling angle. Then clamp the guide block to the workpiece and drill.

Quick Tip: You can eliminate any tendency the guide block has to creep or slide when drilling by gluing a strip of sandpaper to its bottom.

Drills & Drill Presses

ANGLED HOLES ON THE DRILL PRESS

Shims The quickest way to drill an angled hole on the drill press in to insert a shim or scrap of wood under one end of the workpiece until the desired angle is reached.

This works fine when accuracy isn't critical and you need to drill only a hole or two. If accuracy is important or if you're drilling lots of holes, use one of the methods below.

Tilt the table On drill presses with tilting tables, most of the tables require you to remove a stop that locks the table at the 90° position. Once it's removed, you can loosen the nut or bolt that secures the table, tilt the table to the desired angle, and tighten the nut.

A word of caution here: Don't rely on any degree markings on the table—they're notoriously inaccurate. Instead, use a sliding bevel or protractor to set the angle up precisely.

Shop-made tilting table If the table on your drill press doesn't tilt (like the one shown here), not to worry. You can make a simple tilting table to handle angled holes (*see page 88* for detailed instructions).

Just attach the shop-made tilting table to the drill-press table, adjust the angle, and clamp the workpiece to it and drill.

Long bits A common drilling challenge occurs when you need to drill a hole that's deeper than the length of your regular drill bits. If you've got the time (and money), you can hop down to the local building center and purchase a longer bit.

These long bits are commonly available in up to 13" lengths, but they can be expensive. If you have a lot of deep holes to drill, this may prove to be a wise investment.

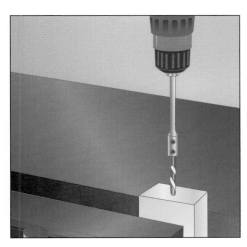

Extensions A more economical alternative to long bits is to buy a drill-bit extension. The only problem with extensions is that they're useful only if the hole you're drilling is larger than ⅜" in diameter—that's the size of the hub that accepts your bits.

The challenge to using either a long bit or an extension is keeping the bit straight as you drill. Here again, sight aids (a try square or a sliding bevel) can help.

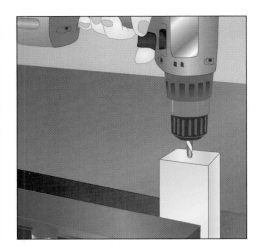

Both directions If you don't have a long drill bit or an extension, another option is to drill from both ends (assuming the workpiece is short enough to allow the holes to meet).

Your best bet for getting the holes to align is to measure carefully and then use a drilling guide or a shop-made guide block similar to one described on *page 59.* Drill slowly and check your progress often.

DEEP HOLES ON THE DRILL PRESS

Tall auxiliary fence Whenever you need to drill a long hole through a workpiece on the drill press, the first thing to do is clamp a tall auxiliary fence to the table (like a flat 2×6).

This fence does two things. First, it supports the workpiece, preventing it from wobbling. Second, it provides a clamping surface for you to clamp the workpiece to and further stabilize it.

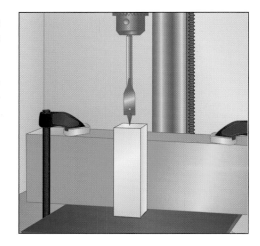

Raise the table Here's a common problem when drilling a long hole on the drill press: The bit is long enough, but the quill stroke isn't deep enough.

The solution is simple. After you've drilled as far as the stroke allows, turn off the drill and retract the quill. Then insert a scrap block under the workpiece. Turn on the drill and continue—the scrap block effectively lengthens the stroke.

Tilt the table to vertical For really deep holes, especially on long pieces, the best way to drill a long hole on the drill press is to tilt the table so it's parallel with the bit. Tilting the table like this provides a solid clamping surface for the workpiece while allowing you to use long bits.

Once you have drilled as far as the stroke allows, you can continue drilling by raising the table. Just make sure to stop periodically and clear out chips.

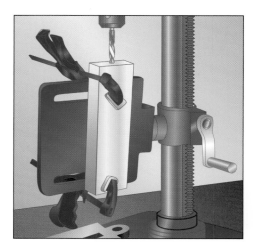

Although not as common as an angled hole or deep hole, the need for a partial hole occasionally pops up in a woodworking project—often as a finger pull in the front of a drawer.

Partial holes also serve as a nice accent on shelf rails and small projects that can benefit from a scalloped edge. Another common use for partial holes is wheel wells for toys like pull-toys, trains, and cars.

The difficulty in drilling a partial hole arises from the fact that there's no support for the centerpoint of the bit as it starts to drill, causing the bit to bow or wander off. The solution is to provide the missing support by clamping a scrap block to the workpiece; *see below.*

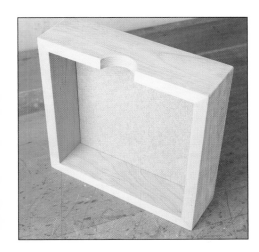

Lip of drawer A partial hole drilled into the edge of a drawer front creates a nice-looking and very functional finger pull. A flush front with no hardware provides a clean look on small jewelry boxes and other such pieces.

This style pull is also useful on small shop storage drawers or other drawers where you don't want any hardware protruding.

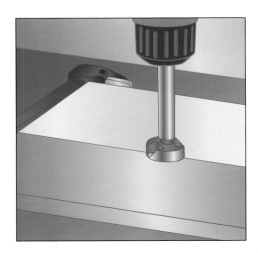

Clamp workpiece to scrap block Although you can drill a partial hole directly on the edge of a workpiece with a Forstner bit, it's best to provide the bit with some additional support. Just clamp a scrap piece to the edge of the workpiece and drill.

The scrap block gives the bit's centerpoint a starting place and will prevent the bit from wandering; the scrap piece also allows you to use drill bits other than Forstner bits.

DRILLING IN ROUND STOCK

V-blocks

If you've ever drilled into round stock, it didn't take long to figure out that you're faced with two distinct problems. The most apparent problem is that the workpiece (whether it's a hardwood dowel or a steel rod) wants to roll around.

A simple shop-made V-block can come to the rescue here and capture that bothersome workpiece; *see below*. Regardless of the type of V-block you make, be sure to clamp both the V-block and the workpiece firmly to the drill-press table (or workbench if you're using a portable drill).

The second problem when drilling in the round occurs as the bit starts to enter the stock. If there isn't an indentation in the surface (made by a centerpunch or an awl), the bit will more than likely wander off.

This is one of the extra steps that really pays off—take the extra time to centerpunch the workpiece before drilling.

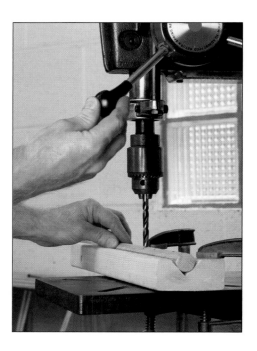

MAKING YOUR OWN V-BLOCKS: TABLE SAW AND PRECUT MOLDINGS

There are two really quick ways to make a V-block for drilling in round stock: on the table saw, and with precut moldings.

If you have access to a table saw, make two intersecting 45° bevel cuts in thick stock (such as a 2×4); *see the drawing at left*.

Or you can make a V-block by nailing a pair of beveled strips (these are 45° trim pieces, available at most home centers and lumber stores) to a scrap wood base; *see the drawing at right*.

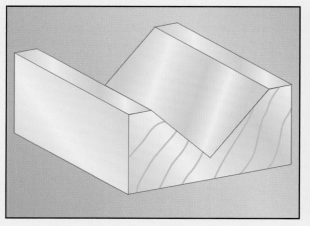

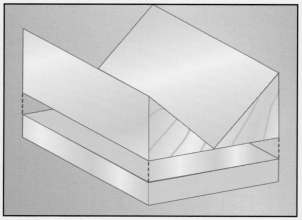

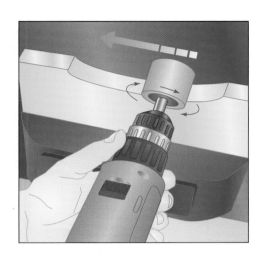

Drum sander in portable drill For sanding curved materials, nothing beats a drum sander. The only trick to using a drum sander is to move the drill in the opposite direction from how the drum is spinning.

If you sand with the rotation, the drum will grab the workpiece and run along the edge, out of control.

Quick Tip: To use the full length of the sanding sleeve, guide the drum gently in and out along the edge as you sand.

Drum sander in drill press There are two major differences when using a drum sander in a drill press instead of a drill. First, you're presenting the workpiece to the drum sander, instead of the drill to the workpiece. Here again, it's important to move the workpiece in the opposite direction the drum is spinning.

Second, since the drum sander can come in contact with the table, there's a gap there, which means you won't be able to sand the full width of a workpiece. To do this, you'll either need to insert the drum sander in a base with a hole in it or flip the workpiece periodically.

Disk sander in portable drill Sanding with a disk sander takes some getting used to. That's because the rotating disk has a tendency to grab the workpiece. The secret is to hold the pad as flat as possible on the surface, use light pressure, and keep the sander moving continuously.

If you stop in one spot, you're guaranteed to leave swirl marks. Even worse, if you're using a coarse grit, the disk is capable of digging in and gouging the work. A firm hand and a light touch are essential to prevent possible damage.

Flap sander in drill press There are a couple of things to be aware of when using a flap sander in a drill press. First, don't exceed the maximum speed recommended by the manufacturer (it's often printed on the side of the wheel).

Second, at high speeds, a flap sander can grab the part and pull you into the wheel. Slow down the drill press and take your time: These sanding wheels, especially when new, can be very aggressive.

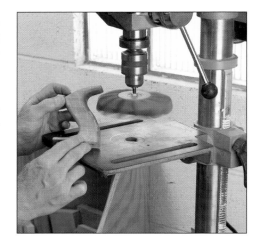

Grinding wheel in portable drill Grinding wheels and points can be used in either a portable drill or the drill press. The only real secret to using them is to let the grinding wheel do the work—take light passes at slow speeds.

Although excessive feed pressure will remove metal more quickly, it'll also most likely remove the temper of the metal, leaving it soft and unable to hold an edge.

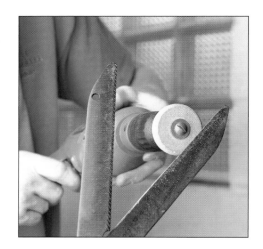

PATTERN SANDING ON THE DRILL PRESS

A pattern sander for the drill press (known by the brand name Robo Sander) works much like a flush-trim router bit. But instead of a bearing and a straight bit, it has a bearing-mounted guide (in this case, a phenolic resin disk) and a sanding drum. In use, a template is carpet-taped to a rough-cut workpiece. Then the template/workpiece sandwich is pressed against the pattern sander. The phenolic guide runs against the template, and the sanding drum shapes the edge of the workpiece to match.

Disk in portable drill By slipping a lamb's-wool pad over a flexible rubber disk, you can put a shine on a vintage automobile or a final luster on a freshly finished piece of furniture.

Automotive buffing compounds make quick work of even the dullest surface when teamed up with a power buffer. When buffing, let the drill do the work and keep the disk in constant motion. It's easy to cut through a wax job or a finish if you stay in one spot too long.

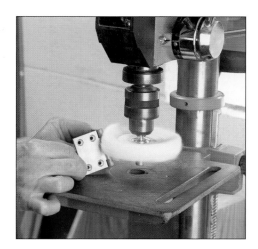

Buffing wheel on drill press You can mount a standard grinder-style buffing wheel on a drill press by making an arbor from a bolt and a nut (or by purchasing an arbor especially for this).

A buffing wheel loaded with the appropriate buffing compound (*see the chart below*) will quickly put a shine on just about anything. Here again, be wary of heavy feed rates, which produce friction and heat—the part will get too hot to handle, and it's easy to melt a finish.

BUFFING COMPOUNDS

Class	Type	Use	Abrasive
Buffing	Green rouge	Stainless steel, steel, brass, aluminum, nickel, chrome	Chrome oxide
Buffing	White rouge	Stainless steel, steel, zinc, aluminum, brass	Calcite alumina
Cutting	Red rouge	Brass, gold, silver	Ferric oxide
Cutting	Gray compound	Steel, stainless steel, nickel, chrome	Aluminum oxide
Cutting	Tripoli	Aluminum, brass, copper, zinc, die casting	Silica

MORTISING ATTACHMENT

A mortising attachment for a drill press can make the tedious job of cutting mortises for joinery quick and accurate. But drilling mortises still takes time. Granted, the mortising drill quickly removes the majority of the wood, but the chisel that shears the sides square uses good old-fashioned elbow grease, supplied by you.

The key to successful mortising is the setup of the attachment. There's nothing complicated about it; you just have to take your time to adjust everything properly. There are couple of things to note.

First, disconnect the power to the drill press before beginning work. Second, make sure to read and follow the manufacturer's directions thoroughly. Third, whenever possible, buy a mortising attachment made by the same manufacturer as your drill press. You can often fit another brand on your drill, but not without some modifications.

1 Attach the chisel holder The first step in adding the mortising attachment is to attach the chisel holder to the drill press. In most cases, the holder attaches to the quill. Lower the quill a couple of inches and lock it in place.

Depending on your mortising attachment, you may need to insert rubber rings between the quill and holder. Most attachments come with an alignment pin to help position the holder.

2 Install and adjust bit Once the holder is in place, you can install the mortising chisel and bit. Insert the chisel in the holder so that the chip-ejection slots are facing the left or right of the drill press (this provides the best ejection and prevents heat buildup).

Then slide the mortising bit up into the chisel so that the upper face of the chisel is approximately $\frac{1}{32}$" below the bottom edge of the holder, and tighten the thumbscrew.

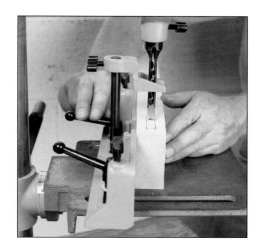

3 Attach and align fence Attach the fence to the drill-press table with the mounting hardware supplied. Then loosen the locking levers and shift the fence to bring the workpiece into the desired position under the mortising chisel.

Next, slide the workpiece along the fence to make sure the mortise will be cut along the desired line. Attach the vertical hold-down to the fence, and adjust it to hold the workpiece with light pressure.

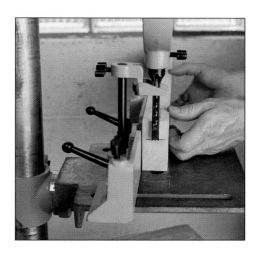

4 Adjust the depth stop Place the workpiece against the fence, and raise the drill-press table until the chisel clears the workpiece by about ½". Then lower the quill to the desired depth and lock the depth stop in position.

If you're cutting through mortises, make sure to insert a scrap block under the workpiece before making the depth adjustments.

5 Cut the mortise Before you cut into your workpiece, take the time to make a test cut first in a scrap block. Double-check the mortise location and depth.

When everything checks out and you're ready to cut a full mortise, start by making the extreme or end cuts. Then take a series of overlapping cuts to remove the waste between the ends. Raise the quill often to aid in chip ejection.

HOLE SAWS

Sawdust relief tip Hole saws are notorious for burning. That's because the teeth are so small and the gullets between the teeth fill up quickly.

Since the sawdust is trapped, it quickly generates friction. Heat builds up and the sawdust starts to burn. You can eliminate this nasty problem completely by drilling a hole inside the perimeter of the cut to allow the sawdust to escape.

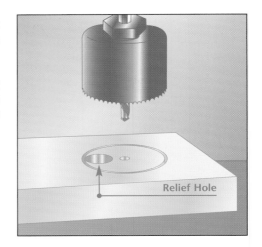

Relief Hole

Keep the kerf clear of sawdust If you don't want to drill a sawdust relief hole, the next best thing to do is stop often and blow out the sawdust that's trapped in the hole saw gullets and in the saw kerf.

Compressed air works best; but if it isn't handy, lung pressure will do. If the hole saw is warm to the touch, let it cool down. If you don't, it'll cut only a short time more before it gets hot enough to burn sawdust.

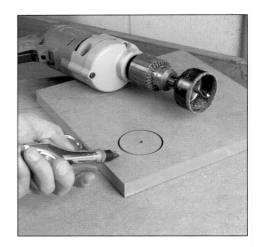

Drill from both directions Just as with any drill bit, a hole saw will cause tear-out when it breaks through the other side. You can prevent this in two ways.

Clamp a backer board behind the workpiece for support, or drill the hole from both directions, as shown. Stopping halfway through also keeps the hole saw cooler by generating less friction, since only half of the cup is buried in the saw kerf.

Of all the drill press accessories that I've seen used, circle cutters get abused the most often. I've wandered into a friend's shop only to be greeted by a cloud of smoke—generated from a circle cutter.

The big problem with these is that although they're designed to cut, they most often end up scraping. That's because they dull quickly—it may not seem like it, but you're removing a lot of wood. You can eliminate half the problem by sharpening the cutters often (I use an oilstone for this).

But the most common mistake I see woodworkers make with a circle cutter is using too high of a speed and/or feed rate. Use your lowest speed, and feed the cutter in slowly. If you detect even the slightest whiff of smoke, your speed and feed rate are too fast.

Finally, just as with a hole saw, cutting halfway through and flipping the workpiece will reduce friction and heat by running less of the cutter in the kerf. A relief hole for sawdust can help, too.

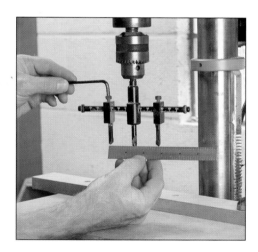

Adjust double wing Adjusting a double-wing cutter can be a challenge. The problem is getting both cutters the exact distance from the arbor. If you don't, one cutter will do the majority of the work and will overheat.

Although you can get them close with a ruler (as shown), the best way I've found to position them accurately is to cut a scrap block as a guide and use it to set both cutters. Just butt it up against each arbor, slide each cutter against it, and tighten.

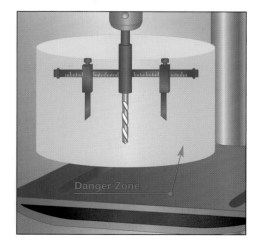

Know the danger zone There's no doubt about it: A circle cutter is a knuckle buster. Since it's spinning and you're concentrating on the cut, it's easy to get whacked inadvertently.

The thing that gets you most often isn't the cutting wings, it's the horizontal shaft they ride on. Try to envision the danger zone as shown, and keep your hands clear.

Quick Tip: It may also help to paint the ends or tips of the horizontal shaft red or bright orange to make them more visible as they spin.

5 USEFUL DRILLING JIGS

Although both the portable drill and the drill press are versatile tools in their own right, their capabilities can be greatly enhanced with shop-made jigs. This chapter features seven jigs, progressing from simple to challenging. Each jig can be made at a fraction of the cost of the manufactured equivalent, using parts and materials common to any well-stocked hardware store or home center.

Since there's nothing quite as frustrating as drilling a hole for a fastener only to find that as you install it, it goes in at some strange angle, I'll start with three simple alignment jigs designed to help you drill a perfectly straight hole with a twist bit or brad-point bit (*pages 80–81*).

Next, if you're into cabinetmaking or woodworking, there's a jig that makes it easy to drill "pocket" holes to join together wood parts, such as the face frame for a cabinet (*pages 82–83*).

This is followed by a nifty jig that allows you to drill straight holes with hex-shank bits or spade bits (*pages 84–85*).

Finally, there is an auxiliary table and fence that will add precision to any drill press (*pages 86–87*), plus a tilting table for tables that don't tilt, and also for those that do tilt (*see pages 88–89* for more on this).

Before you go to make any jig, read through the information on jig materials *on the opposite page.*

For any jig to be worth its weight (and your time), it has to be able to perform its designated function repeatedly and with accuracy.

Both of these start with the right materials. It would be a shame to waste an afternoon building a jig only to find out that it doesn't work as intended because the materials used are inferior.

Even something as simple as applying a finish to a jig can affect its future performance (*see the sidebar on the opposite page*). Note: If you can't find any one of the materials used here, ask your local lumberyard for a source. Odds are, they'll be able to special order it for you.

Jig materials: MDF, hardboard, plywood, and hardwood
There are four materials I use for jigs *shown from bottom to top*: MDF (medium-density fiberboard), hardboard, ¾" and ½" European birch plywood (usually referred to by its common trade name, Baltic Birch), and hardwood.

I use MDF because it's flat and doesn't warp or twist. Most of the fine attributes of MDF also apply to hardboard (often referred to by its common trade name, Masonite). The big difference is that hardboard comes in thinner sheets: ¼" and ⅛". This makes it the perfect material for slides or glides that fit into a kerf left by a saw blade or a router bit.

Types of plywood There are numerous types of plywood available for building jigs. Shown in the drawing *clockwise form top left* are: veneer-core, lumber-core, particleboard-core, and MDF-core. But my favorite plywood material for building jigs is Baltic Birch plywood (*see the photo above*).

The difference between Baltic Birch plywood and cabinet-grade plywood of the same thickness is the number of plies (veneers). Baltic Birch plywood has a lot more plies than cabinet-grade plywood (12 versus 5). More plies means dimensional stability, greater mechanical strength, and better fastener-holding power. This makes Baltic Birch a top choice for the base or foundation of a jig.

FINISHING JIGS

Over the years, I've taken a lot of ribbing about putting a finish on the jigs I build. "Boy, you must be really proud of that, to lavish so much attention on the finish. I mean, it's just a jig." They're right, it is just a jig, but when I take the time to build one, I build it to last. And another way to make sure it lasts is to protect it with a finish—I usually use a couple coats of satin polyurethane. In addition to helping keep it free from shop dirt and grime, this also helps seal the wood parts against moisture, which can cause unwanted movement. So the next time somebody gives you guff, tell them you're adding a coat of precision to your jig (and that yes, you are proud of it!).

ALIGNMENT JIGS FOR A PORTABLE DRILL

Here are three simple alignment jigs you can make for your portable drill that will add a level of precision to your drilling: a drill bracket, a guide block, and a saddle for drilling into 2-by material. Each of these can be made from scraps lying around the workshop. No fancy hardware is required.

Drill bracket

The simplest alignment jig you can make for a portable drill is basically an L-shaped bracket with a notch for the drill bit to rest in (*photo at right*).

The drill bracket consists of two pieces: a ¼" hardboard base and a ¾"-thick hardwood drill support

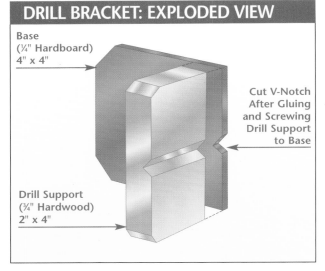

DRILL BRACKET: EXPLODED VIEW

Base
(¼" Hardboard)
4" x 4"

Cut V-Notch
After Gluing
and Screwing
Drill Support
to Base

Drill Support
(¾" Hardwood)
2" x 4"

with a V-notch cut in it; see the Exploded View *below left*.

After cutting the pieces to size, glue and screw them together, then cut the V-notch. You can cut the notch carefully with a hand saw, or on the table saw with the blade tilted to 45°. The way I prefer, though, is to rout the notch with a V-groove bit on the router table.

Quick Tip: After the V-groove is cut, glue a piece of sandpaper to the bottom of the hardboard to help prevent the jig from sliding around during use.

To drill at exactly 90°, start by setting the tip of your drill bit on the mark where you plan to drill. Then slide the drill bracket up so the V-notch cradles the bit. Press firmly down on the hardboard with one hand, and keep the drill bit resting in the V-notch of the drill bracket as you drill the hole.

Guide block

Another jig that's especially useful for shorter, smaller-diameter twist bits is a stepped guide block; *see photo at right*. Each of the seven steps is sized for a different bit, ranging from ¹⁄₁₆" to ¼", increasing in ¹⁄₃₂" increments. To create the seven steps, I cut strips of hardboard and offset them to create the steps; *see the Exploded View on the top of the opposite page*.

By cutting the strips long enough (12⅛") and offseting each strip by 1½", you'll end up with two guide blocks when you cut the blank in half. Once this is done, drill a hole in each of the steps, starting with the smallest bit at the low end and finishing with the largest bit at the other. For 90° accuracy, these holes should be drilled on a drill press. Here again, sandpaper glued to the bottom will help prevent the jig from sliding around in use.

GUIDE BLOCK: EXPLODED VIEW

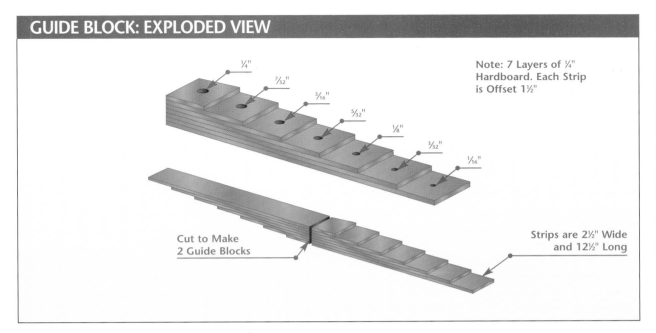

¼"
⁷⁄₃₂"
³⁄₁₆"
⁵⁄₃₂"
⅛"
³⁄₃₂"
¹⁄₁₆"

Note: 7 Layers of ¼"
Hardboard. Each Strip
is Offset 1½"

Cut to Make
2 Guide Blocks

Strips are 2½" Wide
and 12½" Long

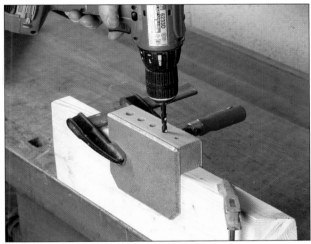

Saddle

The final alignment jig is designed specifically for those times when you need to drill into "2-by" material. The jig is a saddle that slips over and grips the sides of any 2×4, 2×6, etc. (*see the photo at left*).

The jig is made up of three parts: a bit guide, and two ¼" hardboard sides (*see the Exploded View below left*). I used a piece of fir 2×4 for the bit guide since it's the same thickness as the piece the saddle needs to slip over. (I'd recommend that you use hardwood for the bit guide if you feel this jig will get a lot of use.)

Once you've cut the pieces to size, drill a series of holes in the bit guide to fit your larger bits (¼" up to ½", in ¹⁄₁₆" increments). Now you can glue the sides to the bit guide (I also removed the sharp bottom corners of the sides and chamfered all the edges).

In use, just slip the saddle over any 2-by material, clamp it in place, and drill.

Quick Tip: To make alignment easier, you may want to draw lines down the side of the saddle to indicate the centerpoints of the holes. This way you can align these with the hole location marks on the workpiece to be drilled.

SADDLE: EXPLODED VIEW

Bit Guide
(1½" x 6" x 1½" Thick)

¼"
⁵⁄₁₆"
⅜"
⁷⁄₁₆"
½"

Side ¼"
Hardboard
(4½" x 6")

POCKET-HOLE JIG FOR A PORTABLE DRILL

Whether you're attaching the top to a table or making a frame-and-panel door, one of the quickest ways to join together wood parts is to use screws. Quite often in these situations, you're faced with joining together wide pieces.

Instead of using long screws (which have a tendency to break under the torque required to drive them in), a handy solution is to use a jig to cut "pockets" in the face of one of the pieces so you can screw the pieces together with standard-length screws. This pocket-hole jig is basically an angled guide block.

Since you'll often need to drill two pockets at a varying distance apart, the jig features two adjustable guides. Bushings inside the guides stand up to the wear and tear of the rotating bit and help ensure that the pocket is drilled with precision.

2-step bit Although you can use this jig using regular twist bits, it may be worth investing in a step drill bit if you plan on drilling a lot of pocket holes.

This special step bit drills both holes simultaneously because it combines two different diameters in one bit. Step drill bits cost around $15 and are available from most woodworking mail-order catalogs.

Drilling pocket holes To use the jig, loosen the wing nut and adjust the width to the desired spacing. Tighten the wing nut and clamp the jig securely to the workpiece, as shown.

If you're planning to drill only a few pocket holes, you can first drill the pocket with a ⅜"-diameter bit. Then once the jig is removed, use a smaller bit to drill the pilot hole for the screw. For a lot of pocket holes, invest in a step bit (*see above*).

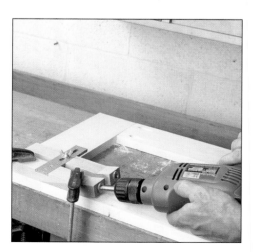

POCKET-HOLE JIG: EXPLODED VIEW

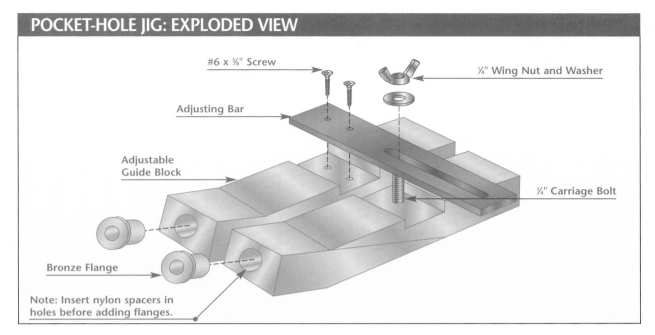

#6 x ⅝" Screw

¼" Wing Nut and Washer

Adjusting Bar

Adjustable Guide Block

¼" Carriage Bolt

Bronze Flange

Note: Insert nylon spacers in holes before adding flanges.

PARTS LIST

Quantity	Part	Dimensions (W×L)	Material
2	Guide blocks	1½" × 7"	¾" hardwood
1	Adjusting bar	1" × 5"	¼" hardboard
2	Flanges	½" OD × ⅜" ID × 1"	bronze
2	Spacers	½" OD × ⅜" ID × 1"	nylon
1	Carriage bolt	¼" × 1"	steel
1	Wing nut	¼"	steel
1	Washer	¼"	steel

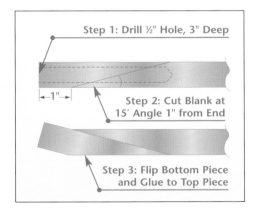

Step 1: Drill ½" Hole, 3" Deep

1"

Step 2: Cut Blank at 15° Angle 1" from End

Step 3: Flip Bottom Piece and Glue to Top Piece

Construction

The challenge to building this jig is creating the long, angled holes and thereby the original "pockets." You could try to drill these holes in thick stock, but getting the angle correct is tricky.

An easier approach is to drill straight holes in ¾" stock, cut the block at an angle, and splice it back together in a configuration that creates the correct angle.

Start with a piece of ¾"-thick hardwood that's 3⅛" wide and 7" long. Then drill two ½"-diameter holes about 3" deep in the blank for a pair of bronze flanges and nylon bushings.

Next cut the blank at a 15° angle, 1" in from the end you drilled the holes in. Once the angle is cut, glue the bottom piece on top of the other.

After the glue is dry, there's one more thing to do before cutting the blank in half. To create a recess for the adjusting bar, cut a 1"-wide, ¼"-deep dado in the top of the blank. Then cut the blank in two lengthwise.

Now all that's left to make is the adjusting bar. It's cut from a piece of ¼" hardboard to fit the recess in the top. A ¼"-wide slot near one end accepts a carriage bolt that passes up through the jig and locks the bar in place by way of a wing nut. The other end of the bar is glued and screwed into place.

SLIDING DRILL GUIDE

You'll never drill another crooked hole with a hex-shank bit or spade bit once you've made this handy jig. For less than $15, you're guaranteed 90° precision.

The sliding drill guide consists of two main assemblies: a bit carriage that slides on a pair of guide rods attached to a base. The bit carriage is made up of a block of wood with two bearings expoxied in it to hold a magnetic bit holder (available at any hardware store). Nylon spacers reduce friction and accurately guide the carriage over the rods in the base.

The base is another small block of hardwood, with a large hole in the center for the bit. The ¼" steel guide rods are expoxied in holes in the base, and a pair of nylon spacers are expoxied in place for additional support.

Using the sliding guide To use the sliding guide, chuck the bit holder in your drill, insert a hex-shank bit in the holder, and drill. Hold the base firmly against the workpiece with one hand and drill with the other.

Quick Tip: To prevent the base from sliding around on the workpiece during drilling, glue a piece of sandpaper to the bottom of the base. Spade bits also work in this jig, but they may need to be shimmed; *see below.*

The drill chuck Since the shank on a spade bit is just slightly smaller than that of a hex-shank bit, you'll need to shim it to fit tightly in the chuck.

I've found that the metal tape sold in hardware stores to seal ducting works well for this (*inset*). There are even chucks available that have a hex shank so that you can drill with any small twist bit.

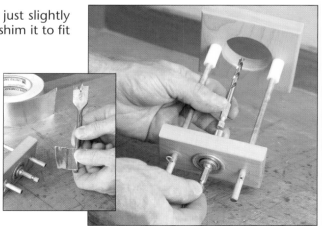

Aligning holes For the carriage to slide smoothly on the guide rods, the holes in the carriage and base must be perfectly aligned. Here's how to do this.

First cut the carriage and base blocks to size, and then attach them together with double-sided tape. Then lay out and drill a ¹⁄₁₆" hole through both blocks for each guide rod and a center hole for the bearings and bit access. After you've drilled the pilot holes, separate the blocks and drill appropriate-diameter holes, as shown in the Exploded View.

Start assembly by cutting the guide rods to length, epoxy them in the base, and add the nylon support spacers.

Next, you can turn your attention to the bit carriage. The heart of this assembly is an ordinary magnetic bit holder. The critical thing here is to find a bit holder with a ³⁄₈" outer diameter (O.D.). This way it will slip right into the radial bearings.

To provide maximum support for the bit holder, I used a radial bearing on both the top and the bottom of the carriage. I found these at the hardware store, but you could try a bearing shop as well (check your local yellow pages under "Bearings"). The fit was so nice on mine that I basically press-fit the bit holder into the bearings. If yours is too loose, just epoxy it in place.

All that's left to do is to epoxy the nylon spacers in the carriage and slip it over the guide rods.

Quick Tip: To provide additional support to the guide rods, epoxy a nylon spacer onto each guide rod before attaching the carriage. Finally, to prevent the carriage from coming off the guide rods, I drilled a ¹⁄₁₆" hole 1" in from the end of each guide rod for a hairpin clip.

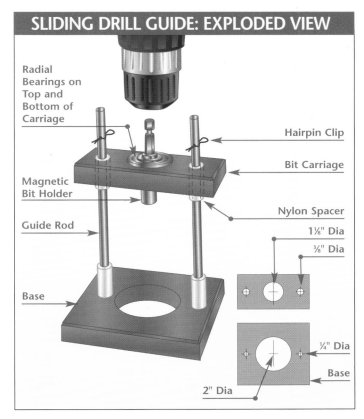

SLIDING DRILL GUIDE: EXPLODED VIEW

Radial Bearings on Top and Bottom of Carriage

Hairpin Clip

Bit Carriage

Magnetic Bit Holder

Nylon Spacer

Guide Rod

1⅛" Dia

⅜" Dia

Base

¼" Dia

Base

2" Dia

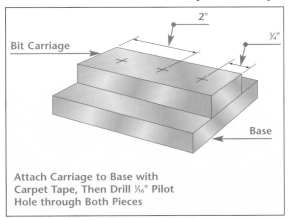

Bit Carriage

2"

¾"

Base

Attach Carriage to Base with Carpet Tape, Then Drill ¹⁄₁₆" Pilot Hole through Both Pieces

PARTS LIST

Quantity	Part	Dimensions (W×L)	Material
1	**Bit carriage**	2" × 4"	¾" hardwood
1	**Base**	4" × 6"	¾" hardwood
4	**Spacers**	½" OD × .257" ID × 1" long	nylon
2	**Guide rods**	¼" × 8"	steel rod
2	**Radial bearings**	1⅛" OD × ⅜" ID	steel
1	**Magnetic bit holder**	⅜" OD	steel
2	**Hairpin clip**	.054" diameter	steel

AUXILIARY FENCE FOR THE DRILL PRESS

One of the simplest ways to improve the accuracy of a drill press is to add an auxiliary table and fence. Without a doubt, the tables on most drill presses are just too small to handle many drilling jobs. An auxiliary table solves that problem by offering a larger surface to support the workpiece.

But that's only half of the precision equation. The other half you'll need is a fence that is easy to position and that locks down reliably. Granted, you can simply use a scrap piece of wood and a couple of clamps, but I always end up fumbling around with that setup.

A better solution is to use a fence that can be adjusted with one hand, like the one shown here. All it takes is a twist of a wing nut to loosen a built-in clamp—then adjust the fence and tighten.

Fence in use For a fence to be useful on the drill-press table, it has to have two features. First, it must be easy to position. Second, it needs a positive locking system.

Both requirements are met with this shop-made table and fence by way of the clamp blocks (*see below*). Just slip the fence in place, and tighten the wing nuts.

Clamp blocks The clamping system consists of a clamp block, a spline, and a bolt and wing nut on each side of the fence (*see the photo*). The secret to this clamping system is the small gap created by a spline that hinges the clamp block to the fence.

The spline rests in kerfs cut in each piece and is taller than the combined depth of the kerfs. This creates a gap that allows the angled clamp block to pull up tight against the matching bevel in the table when the wing nut on the carriage bolt is tightened.

AUXILIARY TABLE AND FENCE: EXPLODED VIEW

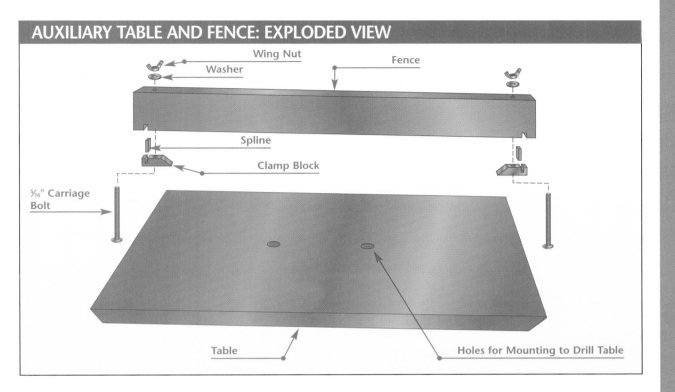

Wing Nut

Washer

Fence

Spline

Clamp Block

⁵⁄₁₆" Carriage Bolt

Table

Holes for Mounting to Drill Table

PARTS LIST

Quantity	Part	Dimensions (W×L)	Material
1	Table	11½" × 20½"	¾" MDF
1	Fence	2½" × 24"	1½" wood
2	Clamp blocks	1½" × 2½"	¾" MDF
2	Splines	¾" × 1½"	⅛" hardboard
2	Carriage bolts	⁵⁄₁₆" × 4"	metal
2	Washers	⁵⁄₁₆"	metal
2	Wing nuts	⁵⁄₁₆"	metal

To make the auxiliary table and fence, start by cutting the table to size; *see the Parts List.* Then cut 45° bevels on each end. Position the table on your drill press so it's centered on the table and far enough forward so the height-adjustment handle clears it when turned.

Clamp the table in place, and mark up through the slots on the drill-press table onto the auxiliary table. Drill holes in the auxiliary table for the mounting hardware of your choice (I used ⁵⁄₁₆" carriage bolts).

Now you can cut the fence to size; just make sure

the wood you use is flat and free from twist. All that's left is to make the clamp blocks. The safest way to do this is cut a 45° bevel on the long edge of 6"-wide by 8"-long piece of MDF. Then rip the beveled end off to a length of 2½", and cut the individual blocks from this blank to a width of 1½".

All that's left is to cut the kerfs for the splines ⁵⁄₁₆" deep, ⅜" in from the ends of both the clamp blocks and the fence. The splines are cut from ⅛" hardboard and are glued in place. Drill holes for the carriage bolts through the fence and stop blocks at the same time to make sure they line up.

TILTING TABLE FOR THE DRILL PRESS

Although this shop-made table is designed for drill presses that don't have a tilting table, it's still useful for those that do.

First, if your table tilts only in one direction, this table can be mounted to drill angled holes in virtually any direction. Second, many tilting tables can tilt only to 45°; this table will allow you to drill angles greater than that.

Finally, drilling compound angled holes is now possible by angling your table and attaching this one perpendicular to it—compound holes are common in chair construction.

The tilting table is just two pieces of ¾" MDF (medium-density fiberboard) held together with a pair of no-mortise hinges (although any stable ¾"-thick material, like plywood, would do). Two friction lid supports allow you to tilt and lock the table to the desired angle.

Table in use The base of the tilting table attaches to your drill press table by way of a set of carriage bolts that pass through the slots in the table. Wing nuts provide the clamping power.

The table is adjusted to the desired angle (*see below*), and the workpiece is clamped to it. For added precision, consider clamping a fence or stop to the bottom of the workpiece, as shown, to prevent it from shifting as you drill.

Adjusting the angle A pair of friction lid supports provide a positive way to lock the table in at the desired location. Loosen the screw on each side until the table pivots freely.

Then set the table to the desired angle, using a sliding bevel or by inserting a scrap cut to the desired angle in between the base and the tilting table. Now just tighten the screws to lock the table in place.

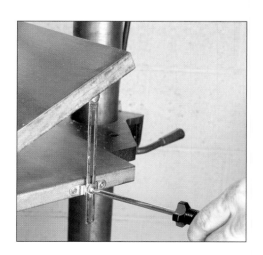

TILTING TABLE: EXPLODED VIEW

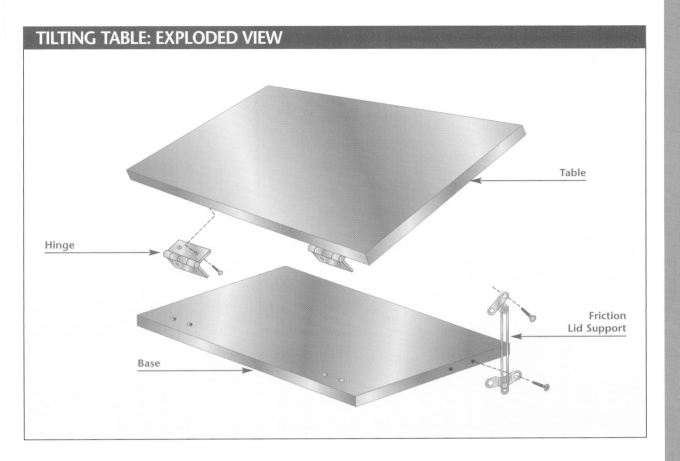

Table

Hinge

Friction
Lid Support

Base

PARTS LIST

Quantity	Part	Dimensions (W×L)	Material
1	Base	10" × 17"	¾" MDF
1	Table	10" × 19"	¾" MDF
2	Hinges	1½" × 2"	Brass
2	Friction lid supports		Brass

6 MAINTENANCE AND REPAIR

With the cost of the average high-quality electric drill rising well over $100 and a drill press costing anywhere from $150 to $750, it makes financial sense to protect your investment with periodic maintenance.

Periodic maintenance will also increase the life and accuracy of your drill or drill press. Simple and easy to perform, the maintenance of your drill or drill press can be broken down into just a few tasks: cleaning and inspection, lubrication, proper storage, and if necessary, repair.

Surprisingly, one of the best ways to maintain a drill or drill press is to buy the right one for the job in the first place (see Chapter 1). For example, I've seen plenty of first-time buyers purchase a weak, high-rpm portable drill and then try to drive screws with it, only to

end up with a burned-out motor or stripped gears.

Even if you maintain, lubricate, and store your drill or drill press properly, you will eventually experience a failed part that will need repair. If you're not comfortable doing the repair work yourself, look in the yellow pages for tool repair shops.

If, on the other hand, you're adventurous, let's take a look at some of the most common drill repair jobs. In order of complexity, we'll tackle replacing: a chuck (pages 94–95), an electrical cord (pages 96–97), brushes (pages 98–99), and finally, a power switch (page 100).

On drill presses, I'll show you how to replace a chuck (page 102), check a quill for runout (pages 103–104), adjust and

replace belts (pages 105–106), and adjust and replace pulleys (page 107).

Any part that I don't cover in this chapter falls into my "don't try to replace this" category. This includes bearings, spindle, and gear assemblies. If you're experiencing a bearing problem or gear assembly–related problem (noisy, grinding operation), take your drill or drill press in to a repair shop. They've got the specialized tools and know-how to get the repair done quickly, safely, and at minimal cost.

Finally, I'll show you how to clean, lubricate, maintain, and sharpen the accessories you use most often: drill bits (pages 108, 109, 110, and 111–113, respectively).

The number one thing you can do to extend the life of your portable drill is to blow it out regularly (with compressed air, if you've got it); *see below.* The chuck is also worthy of some special cleaning attention. Spin the chuck jaws to the fully retracted position, and blow out dust and debris. Close the jaws about halfway and repeat. Continue this procedure until the chuck jaws open and close smoothly without binding.

After you've blown out all the dust and dirt form the drill, you can clean it with mild soap and a damp cloth. Just make sure that no fluids enter the drill housing.

Do not use solvents of any kind. Solvents will degrade the plastic of both the drill housing and the electrical cord and will certainly cause problems in the future.

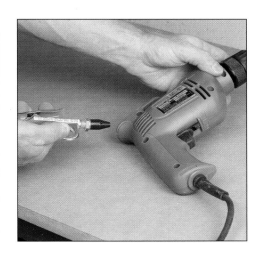

Blow out debris As you blow out the drill with compressed air, pay particular attention to the vents and switches. Dust and debris that get inside will cake the motor housing and electrical contacts, causing them to overheat, work intermittently, or just downright fail.

This is particularly important when you're working on jobs that generate fine or abrasive dust, such as drilling into drywall or masonry, or when removing an old finish with a disk sander or abrasive wheel.

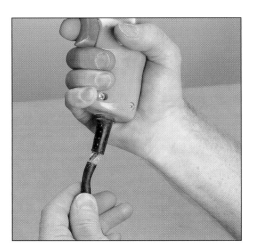

Inspect for wear Another good habit to develop is to inspect your drill before you use it. This not only helps maintain the life of the tool, but also protects you from working with an unsafe tool. Look for loose screws, misalignment, binding or moving parts, broken parts, or damage to the cord.

When you power up the drill, pay attention to how it feels and sounds. If you notice any odd vibration or noise, stop at once. Don't use the drill until it has been checked out and repaired.

LUBRICATION

Unless you're working a portable drill hard (such as installing decks or drilling into masonry a lot), lubrication on a regular basis is not necessary. Most drills have sealed bearings that enjoy a long life.

If you are working the drill hard, it may be necessary to drop it off at a repair center for lubrication. How often you need to do this will depend on the type and amount of work you do.

Most manufacturers suggest that at least once a year, you should have the drill lubricated and the brushes checked. (There will be more on brushes later.)

Drill presses, on the other hand, do require periodic attention. The main reason for this is the environment that they're often in—a dusty workshop. Sawdust, metal shavings, and dirt will cling to the lubricated parts of a drill press. So it's important to clean and lubricate these parts regularly; in particular, the chuck, the column, and the quill.

1 Access the chuck One drill part that can use periodic lubrication is the chuck. On a drill press, the best way to lubricate the chuck is to first remove it, as shown (*see page 94*).

On a portable drill, position the unplugged drill (or cordless drill with the batteries removed) so that the chuck is pointing up and the jaws are fully retracted.

2 Lubricate For a drill-press chuck, hold it upside down and let gravity pull some light machine oil down into the gearing for the jaws. Rotate the exterior ring through its full range of motion a couple of times to help work the oil into the gears. Wipe off any excess oil before reinstalling it.

With a portable drill, apply a drop or two of light machine oil to each of the chuck jaws. (Any more than this, and you run the risk of the bit slipping or of spraying yourself with oil as you power up the drill.) Here again, run the outer ring through its complete range of motion a couple of times, opening and closing the jaws to help spread the oil to the gearing.

3 Column If your drill press doesn't use a rack-and-pinion system to raise and lower the table, it's imperative that the column be clean and well lubricated. Start by running an abrasive pad up and down the column to remove any old grease, sap, or bits of sawdust. Then clean the column with acetone or mineral spirits.

When dry, apply a generous coat of paste wax. Let it dry, and then buff it out both to retard rust and to allow for smooth, easy table adjustment.

4 Quill The quill on a drill press also requires periodic lubrication. Start by lowering the quill and locking it in place. Wipe off the quill with a clean rag dipped in acetone or mineral spirits.

For a lubricant, I prefer the spray type, as it is easier to apply it with precision. A light coat is all you'll need. If you do apply a bit too much, wipe it off immediately—the last thing you want is lubricant dripping down onto a workpiece and ruining it.

5 Rack-and-pinion In any woodshop, the rack-and-pinion mechanism on a drill press will slowly acquire a coat of sawdust. This will mix in with the lubricant to form a gooey paste that will eventually make the table difficult to adjust.

To prevent this problem, periodically blow off as much dust as possible, using compressed air. If you notice a buildup, scrub away the old goo with an old toothbrush. Then apply a fresh coat of white lithium grease to the rack with a clean toothbrush or rag.

REPLACING A CHUCK

1 Remove screw Your old chuck has given up the ghost, or you've decided to upgrade to a keyless chuck. The only challenge is removing the old chuck.

To do this, first unplug the drill (or remove its batteries), and then retract the jaws of the chuck so they're fully open. Then use a screwdriver (or an Allen wrench, depending on the screw head) to loosen the screw that holds the chuck onto the spindle. This screw will be a left-handed thread, so turn it clockwise to loosen it.

2 Loosen chuck Since a drill is subjected to a lot of vibration over its lifetime, the chuck develops a very tight grip on the threads of the motor spindle. I've replaced a lot of chucks, and I've never encountered one that didn't need a little friendly persuasion to give up its grip.

One way to do this is to insert the key into one of the holes in the chuck. Then give the key a sharp rap to break the chuck free. If it's stubborn, apply penetrating oil, wait 15 minutes, and try again.

KEYLESS CHUCK FOR THE DRILL PRESS

I've gotten so spoiled by the keyless chuck on my cordless drill that I started to get impatient with the keyed chuck on my drill press. I just assumed that no one made a keyless chuck for the drill press.

Needless to say, I was pleasantly suprised when I came across one in a tool catalog. Excited, I ordered one and have been very happy with it. Keyless drill-press chucks come in various sizes and tapers. Contact Grizzly Industrial, Inc. at www.grizzly.com for more information.

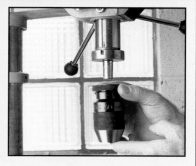

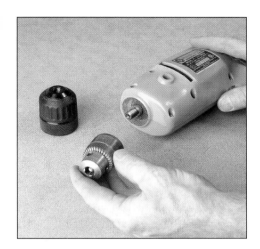

3 Remove the old chuck Once the chuck is loose, unthread it the rest of the way by hand. Take the time now to inspect the motor spindle.

If it's dirty or rusty, clean the threads with a brass brush. Then apply a few drops of light machine oil to the threads and wipe off the excess with a clean rag.

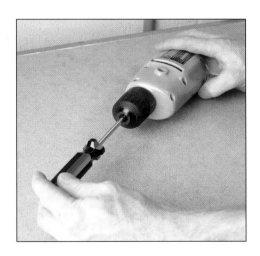

4 Install the new chuck All that's left is to install the new chuck. Thread on the new chuck by hand until it bottoms out. Then reinstall the left-handed spindle screw. Don't over-tighten this—it's not necessary; normal use will spin it tighter.

Quick Tip: If you've got a well-stocked hardware store nearby, consider replacing this screw with an Allen-head screw—it'll be a lot easier to remove the next time.

REMOVING A KEYLESS CHUCK

The standard way to remove an old worn-out chuck is to insert the key in one of the holes in the chuck and give it a sharp rap (*see Step 2 on page 94*). But what if the chuck you're replacing is a keyless chuck? The solution is simple: Just insert the short end of a sturdy Allen wrench (¼" or larger) into the chuck and tighten. Now, holding the drill firmly against a workbench, strike the Allen wrench firmly with a mallet to loosen the chuck.

REPLACING A CORD

Field change Probably one of the simplest repairs to do, replacing an electrical cord is often the most neglected—that is, until someone gets shocked. Worn or missing insulation (particularly where the cord enters the drill) is unsafe, and the cord should be replaced immediately.

The ease with which you can replace an electrical cord depends primarily on the manufacturer of the drill. Some manufacturers have designed the drill so this is simplicity itself. The cord on the drill shown here twists and pulls out for super-easy replacement.

Access panel Other portable-drill manufacturers offer detachable handle housings that provide relatively open access. These are usually held in place with only a couple of screws.

Quick Tip: Regardless of ease of access, once the ends of the electrical cord are in sight, make a note of wire colors, locations, and wire routing.

Split case Some drills require splitting the halves of the case apart for access. These are the least friendly for user maintenance.

Be warned that it's a lot easier to open a split-case drill than it is to reassemble it. That's because the wiring is routed through notches in the case and can easily be pinched if you're not careful; you may be better off taking this type of drill to a service center.

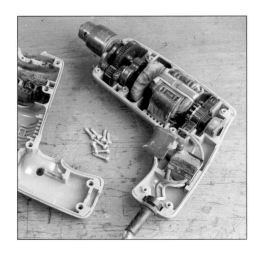

Replace cord How the wires of the cord are connected inside the drill will depend on the manufacturer. There are three common variations: The wires can be screwed in place, soldered, or inserted into wire clips or holders (this is the most often seen in newer drills).

If the wires on your drill are screwed or soldered in place, a screwdriver or soldering iron/gun will get the job done. If they're the newer clip-type, see the sidebar *below*. After you've replaced the cord, be particularly careful to route the wires so that they don't get pinched when you reassemble the case.

DEALING WITH MYSTERY WIRING

At first glance, it doesn't appear that the wires going into switches and other internal electrical parts on a drill can be replaced without discombobulating the entire thing. Fear not. These wires are held in place with a clip inside the switch. All it takes to release the wire is the right tool: in this case, a dental pick or probe used for assembly/disassembly work (a small nail or a thin awl will also get the job done). I found a four-piece probe set at Radio Shack (catalog number 64-1941) for around $10. To release a wire from the switch, take a straight probe and insert it down alongside the wire. This should release the clip, allowing you to pull the wire out. The new wire can usually be re-inserted directly into the hole without using the probe (stranded wire is best inserted with the probe to prevent unraveling).

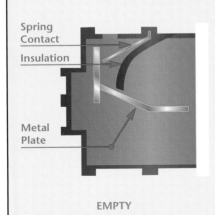

Spring
Contact

Insulation

Metal
Plate

EMPTY

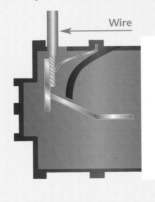

Wire

WIRE INSERTED

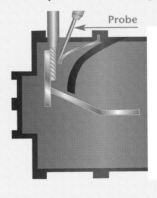

Probe

WIRE REMOVAL

REPLACING BRUSHES

The brushes in a portable drill motor provide a way to transfer electrical current to a rotating object (in this case, the armature). Brushes are made up of highly conductive carbon particles pressed together in a small rectangular bar.

One end of the brush is curved to match the diameter of the armature. A spring inserted between an end cap/wire assembly and the brush pushes the brush against the armature. By the very nature of this pressing and rubbing action, the brushes will wear down over time.

As the brushes wear and approach the end of their life, you may notice a decrease in power and an accompanying shower of sparks. If even one brush goes completely bad, the motor will stop. Both of these situations call for replacing the brushes. When you go to do this, keep in mind that you should always replace brushes in pairs.

Easy access Just as with replacing a power cord, the ease with which you can access brushes depends primarily on the manufacturer. The manufacturer of the drill shown provides external caps that when unscrewed provide instant access to the brushes. Another manufacturer incorporates a brush carriage on many of their drills that pops out of the side of the drill housing. The carriage even holds a pair of extra brushes and a replacement screw.

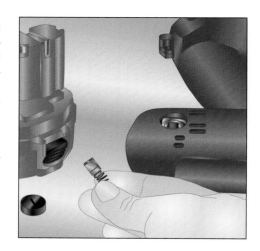

Note: On some of the larger, more industrial drills (particularly impact drills), manufacturers have added a built-in service reminder light that, when lit, indicates that there are only so many hours of use left before the brushes fail.

CRACKING A CASE

In my experience, opening up a corded drill to access brushes is like popping the hood on a 1950s car. Nothing fancy; everything is accessible and for the most part understandable. Cracking the case on a cordless drill, however, is like opening up a fuel-injected, emission-control–laden car of today.

That's assuming you can get into the drill in the first place. Many manufacturers assemble their products with small Torx-head screws to make it clear that they don't want you inside. And there's a very good reason for this: Most cordless drills have advanced electronic circuitry that is static sensitive; that is, a spark generated simply by walking across a carpeted room can easily damage these components.

When manufacturers say that there are no user-serviceable parts inside, they mean it. Usually, the brushes are sealed within the motor housing. This is one of the situations where you're better off bringing in your cordless drill to an authorized repair center (*see page 99*). The bottom line on a cordless drill is, if the drill stops working, try a fresh battery. If the problem persists, take it in. **Safety Note:** If your battery is the problem, make sure that you recycle it; the law prohibits any other method of disposal.

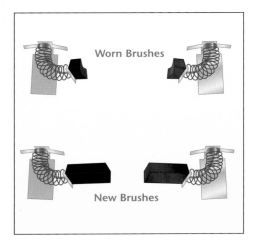

Worn Brushes

New Brushes

Worn brushes If you think you have a brush problem or you'd just like to check them for wear, remove the brushes and do a visual inspection. There are three things to look for: the condition of the brushes, their length, and the spring tension (*see below* for more on spring tension).

In terms of condition, what you're looking for is a nice even gloss on the end of the brush. If the brush is scarred, it needs to be replaced. As to the length of the brush, it's difficult to know when to change brushes unless you know how long they were to start with. As a general rule of thumb, if you've got less than ¼" left in length, replace them. Compare the worn set in the top of the drawing with the new replacment set on the bottom.

New Spring

Old Spring

Spring tension Even if the brushes are in good shape and they're long enough, they won't be worth a darn if there isn't sufficient spring tension. Note the difference between the new spring in the top of the drawing with the worn spring on the bottom. If there isn't enough pressure exerted by the springs, the brushes will make only intermittent contact and your drill will operate sluggishly. If in doubt, replace them.

Finally, if you find that your brushes are wearing unevenly or you're replacing them too often, do a visual inspection of the armature. If one or more of the segments is bad, it can cause excessive brush wear and tear. If it's worn, see a service center for a replacement armature.

SELECTING A REPAIR CENTER

Your drill or drill press stopped working. You page through your owner's manual and find the list of service centers. There's one nearby, and the sign on the door says "Authorized Service Center." What exactly does that mean? In most cases, it indicates the owner of the shop has signed a contract with the manufacturer to perform in- and out-of-warranty repair work on their products. Does the owner have to do anything special to become authorized? Most tool companies check out the owner as if they were planning on hiring him. Once satisfied, the best companies provide tool-specific training, technical data, full access to parts, and service manuals.

Does this mean that the shop is qualified to repair your tool? Not necessarily. It pays to check references ahead of time with other tool users. If you hear good things, talk to the owner. Get a written estimate. If he doesn't instill confidence or the estimate seems out of line, try another shop or call the manufacturer for a recommendation.

REPLACING A POWER SWITCH

The complexity of replacing a power switch on a portable drill will depend on the manufacturer and on the type of switch you're replacing. If it's just a simple on/off switch, it's an easy task.

However, if the power switch is a variable-speed switch and/or there's a built-in forward/reverse switch, it can be quite a challenge. Your first order of business in either case is to look in your owner's manual for a parts list and order a new switch (from the manufacturer or from a local repair shop).

There are two basic ways that wires are connected to power switches—screw and clip—so you'll need either a screwdriver or a probe (*see the sidebar on page 97*). As when replacing an electrical cord, it's a good idea to note wire colors and locations before removing the old power switch. And as always, be careful how you route the wires before you re-assemble the case.

Simple on/off switch Simple on/off switches make and break a connection for a single wire. There are two connections on the switch—just like a single-pole electrical switch.

Before attempting any repair, unplug the drill or remove the batteries. Then disconnect each of the wires from the switch and connect each in turn to the replacement.

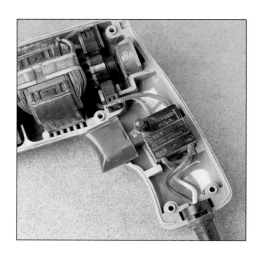

Variable-speed switch Variable-speed switches typically have multiple wires running into them—it all depends on the manufacturer. The most reliable way to replace a variable-speed switch is to disconnect one wire at a time and connect it to the corresponding terminal on the replacement switch.

Pay particular attention to wire routing, as some wires often run under the switch and can easily be pinched when the case is reassembled.

STORING PORTABLE DRILLS

One area of maintenance that is often overlooked is storage—not only where you store your drill, but also how it's stored. Something as simple as wrapping the cord can appreciably extend the life of a drill; *see below.*

Naturally, extremes in temperature are not good for your drill. Heat will break down some plastics, including the cord. Excessively high temperatures (above 100° Fahrenheit) can cut the life of a tool dramatically—as much as in half. And the other extreme, cold, can cause its share of problems.

Repeated freeze/thaw cycles (such as in a garage in winter) are tough on tools—particularly on lubricants and batteries. Batteries should be stored indoors at all times. Note: Just as you can increase the life of regular batteries by storing them in a refrigerator, you can extend life on a Ni-cad (nickel-cadmium) battery the same way.

1 Location If you own a cordless drill, store the drill in the case (if it came with one) or set aside a spot in the shop where it's clean, cool, dry, and uncluttered (good luck finding one of those!).

If you're working in a garage in the wintertime, don't forget to bring in the battery after a day in the shop. A frozen battery is an expensive paperweight. Corded drills will also benefit from a storage location as described above.

2 Wrap cord How you store your drill is also important. If you own a corded drill, the manner in which you wrap the cord can significantly decrease or increase the life of the tool. What you don't want to do is wrap the cord around the tool; this puts too much pressure on the cord protector, where the cord enters the drill.

The correct way to wrap a cord is to form a series of loops and then wrap the plug end around these loops, as shown. This method prevents breaks in the cord, particularly those prone to happen near the housing.

Drills & Drill Presses

REPLACING A DRILL-PRESS CHUCK

Over time, the chuck on a drill press can wear out: The bearings give up, the jaws just don't hold bits secure anymore, or the teeth on the outer ring that you use with the chuck key get chipped, broken, or stripped. Another reason to replace a chuck on a drill press is to upgrade to a keyless chuck; *see the sidebar on page 94* for more on this.

Although you might think replacing a chuck on a drill press would be very similar to replacing one on a portable drill, it's not. There are a couple of major differences.

First, the chuck doesn't thread onto the motor spindle—it's a press-fit (*see below*). Second, there's no screw inside the chuck to hold it in place. And third, because a drill press is inherently more precise than a portable drill, there's more involved with setup and adjustment (*see pages 103–104*).

1 Remove old chuck The mounting hole in a drill-press chuck has a slight taper that matches the taper on the drill-press spindle. The chuck is press-fit onto this taper, much like a drive center fits into a lathe's headstock.

To remove the chuck, first lower the quill and lock it in place with the quill lock. Then position a scrap of hardwood on the top edge of the drill chuck and give it a sharp rap with a hammer. Rotate the chuck a quarter-turn and then strike it again. Continue like this until the chuck pops off.

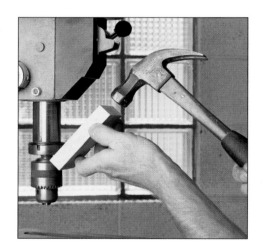

2 Press on new chuck Before you press on the new chuck, check the spindle for dirt, grease, or burrs. Remove any you find with solvent and a small mill file.

The new chuck can be press-fit onto the spindle by using the quill. Simply press the chuck in place and then, with the table locked in place directly beneath it, lower the quill to press it on. This doesn't require Herculean strength, just firm downward pressure.

As I've mentioned previously, a drill press is designed to drill accurate and precise holes. In order for this to happen, the quill must be true and the chuck mounted perfectly straight on the spindle. If either one of these is out of alignment—a condition referred to as runout—the bit will wobble, and starting a hole at a precise point will be almost impossible.

You can check for runout with a dial indicator (*see Step 1 below*) or use an inexpensive low-tech ver-

sion using a scrap block of wood and a feeler gauge (*see Step 2 below*). If the runout is caused by an improperly seated chuck, it's easy to fix; *see Step 4 on page 104*. If the quill is bent, it'll need to be replaced.

At the same time you check for runout, it's a good idea to check the quill for excessive play (*see Step 3 on page 104*). And finally, if you're after absolute precision, see Step 5 on *page 104*.

1 Check runout with a dial indicator The most reliable and accurate way to check your quill for runout is to use a dial indicator like the one shown.

Unplug the drill press, and then butt the probe of the dial indicator up against the chuck. Turn the motor pulley or belt by hand to rotate the chuck. Movement in excess of 0.005" indicates excessive runout.

2 Check runout with a block and feeler gauge If you don't have a dial indicator, you can still check for runout with this low-tech method.

Unplug the drill press and insert a large-diameter bit. Raise the table up so about 1" of the shank is exposed. Then press a scrap of wood against the shank. Rotate the motor pulley or belt by hand. Any runout will force the block away from the shank. Measure the largest gap with a feeler gauge; here again, 0.005" is excessive.

3 Testing for excessive play In addition to a bent spindle, excessive runout can be caused by worn-out bearings. To check for this, first lower the quill as far as it will go and lock it in place with the quill lock. Then grip the quill, as shown, and try to move it from side to side.

If the bearings are okay, there should be no movement at all. If you can shift the quill, the bearings need to be replaced—a service center is your best bet for this job.

4 Adjusting runout An improperly seated drill chuck is often the cause of runout. To correct the problem, tighten a large-diameter drill bit into the chuck. Strike the shank of the bit with a hammer on the side that runs out—that is, the side that caused the scrap block or dial indicator to move the most.

After you've adjusted the chuck, check the runout again, and repeat as necessary. If the runout is still excessive, the spindle is likely bent and needs to be replaced.

5 Alignment marks If a drilling job requires absolute precision, you can check for and adjust for runout with the bit you're planning on using.

If you're likely to repeat the job in the future—say, for instance, you're reboring the arbor hole on saw blades—then you can save the setup time by making a set of scribe marks on both the chuck and the drill bit. Then the next time you need to drill with this bit, all you have to do is align the marks for rock-solid precision.

If you've noticed that your drill press isn't running as smooth as it used to, you may have a belt problem. A belt that's worn, stretched, or cracked can cause the drill press to vibrate. Also, if the step pulleys in which the V-belts ride are not aligned, it can also cause vibration and will eventually lead to excessive belt wear and tear.

Fortunately, adjusting and replacing belts are fairly straightforward. Releasing belt tension is usually just a matter of loosening a single thumbscrew or

knob. Inspecting a belt for wear takes only a few seconds, and replacing a belt is as easy as changing speeds (*see pages 46–47*)—the only critical thing is that you get an exact replacement.

If you've got a number of belt-driven machines in your shop, consider purchasing a couple feet of interlocking belt; *see the sidebar on page 106.* This style belt uses links that can be interlocked to form any size belt you need—sort of a "universal" belt.

1 Check for wear To check a V-belt for wear, unplug the drill press and release the belt tension. Then lift up the belt/pulley cover and gently lift off one belt at a time.

Turn the belt inside out and inspect for cracks, excessive wear, and rubber that's degrading (indicated typically by rubber that's crumbing or loose). Bend the belt around its entire perimeter to check the entire surface. If you notice any problems, replace the belt; *see the sidebar on page 106* for options on new belts.

CHECKING BELT TENSION

If you notice that your bits often get stuck or seem to bog down, the belt tension might not be adjusted properly and the belt is slipping. Belt tension is adjusted on most models by loosening a thumbscrew and forcing the motor and pulley father away from the other step pulleys. Check for proper tension with the drill press unplugged by squeezing the belt together in the center, as shown. A combined deflection greater than 1" will result in belt slippage.

Drills & Drill Presses

105

2 **Check alignment** If you do discover that your belts are worn or that they tend to wear quickly, your step pulleys may be out of alignment—one is higher than the other. When this occurs, the edge of a belt will rub against the steps on the pulley, causing excessive wear.

You can check pulley alignment by placing a straightedge across the pulleys. The straightedge should lie flat on both pulleys. If it doesn't, loosen a pulley setscrew and raise or lower it as needed.

3 **Install new belt** Replacing a belt is simple. After you've removed the old belt, slip one end of the new belt over one step pulley. Then slip the other end over the appropriate step on the opposite pulley. Rotate the pulley by hand to ease it into place.

You can find V-belts of various sizes at any automotive store (or you may have to special-order it from the manufacturer). Take your old belt with you to make sure you get the right size replacement.

BELT TYPES: V-BELT AND LINK

There are two basic types of belts that drive drill press spindles: V-belts (*left*) and an interlocking style (*right*). V-belts come in a variety of thickness to fit different step pulleys. They also come in various diameters to span a wide range of distances between pulleys. Although more expensive than standard V-belts, interlocking belts are sold by the foot and offer the advantage of custom sizing; by adding or subtracting links, you can make any belt diameter you need.

ADJUSTING AND REPLACING PULLEYS

Although they don't often need maintenance, the step pulleys in a drill press can make the difference between a smooth-running machine and a belt-gobbling beast. If the pulleys are out of alignment (*see page 106*), they can cause the drill press to vibrate and will greatly reduce the life of your belts. Pulleys that have the edge of their steps cracked, broken, or nicked can also lead to premature failure of a belt.

Aligning or replacing a pulley is simple, in theory: You just loosen a setscrew and adjust or lift off the pulley. But most pulleys have been pressed on in the factory and over the years develop a death grip on the spindle due to the vibration caused by day-to-day use.

The solution is to borrow a specialty tool used in the automotive trades: a bearing puller. A large bearing puller will usually free even the most stubborn pulley. A word of caution here: Bearing pullers are capable of exerting a lot of force, so take it slow and wear eye protection.

1 Remove setscrew Most step pulleys are held in place on a spindle by way of a setscrew that often rests against a flat on the spindle shaft. Loosening and backing out the setscrew with an Allen wrench or screwdriver will allow you to adjust the pulley up or down, or remove it.

In cases where a step pulley is pressed onto the spindle, you'll likely need to use a large bearing puller to remove it; *see Step 2.*

2 Remove pulley with a bearing puller Loosening a setscrew doesn't mean that a step pulley will automatically behave and pull off easily.

In most cases a little encouragement is needed in the form of a bearing puller. If possible, use the type that has three arms—they spread the force more evenly on the pulley. Loop the arms over one of the steps, and engage the threaded stud in the dimple in the spindle. Slow, steady pressure works best; penetrating oil can help free a stubborn pulley.

Drills & Drill Presses

CLEANING DRILL BITS

1 Brass brush The best all-around tool I've found for keeping drill bits clean is a small toothbrush-style brass brush. The brass fibers won't scratch the metal, but they are stiff enough to remove most of the common gunk found on drill bits: sawdust, metal shavings, resin, and dirt.

Pay particular attention to the flutes and the cutting edges. In cases where resin has built up, a shot of commercial resin remover followed by a brisk scrubbing will do the job.

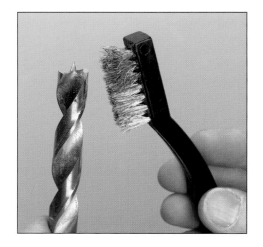

2 Clean rag After you've scrubbed a bit clean with a brass brush, use a clean rag to wipe it off. Here again, if you notice resin buildup or grime, wet the rag with resin remover, acetone, or lacquer thinner, and rub the bit clean.

Allow the bit to thoroughly dry before applying any type of lubrication (*see page 109*).

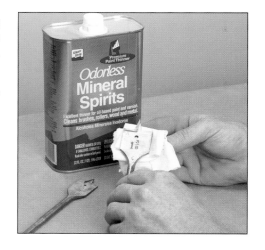

3 Degrease masonry bits If you're drilling with SDS or spline-shank bits in a hammer drill, it's a good idea to degrease the shank after each use. If you don't, the grease can harden and will interfere with the gripping mechanism in the chuck the next time the bit is used.

To do this, dampen a rag with acetone and wipe the shank of the bit. A cotton swab (such as a Q-Tip) will reach into any recesses to remove old grease and grime.

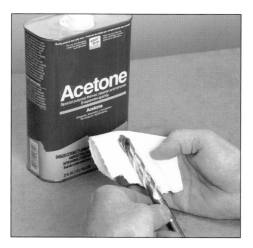

Keeping your bits clean and lubricated is one of the simplest ways to increase their life span. Obviously, you don't want drill bits dripping with oil, especially when you're working with wood: The oil will stain the wood and can cause finishing problems.

Instead, a thin coat of light machine oil or one of the many spray lubricants (especially those that dry, leaving a thin layer of protection and lubrication) is what works best here. Taking care of your drill bits is one of those nice puttering-around-the-shop kind of jobs to do on a cold winter day.

In addition to lubrication, proper bit storage will also increase bit life. Whenever possible, keep drill bits in the jackets, sleeves, or the cases they came in (see page 31). One of the quickest ways to damage bits (particularly those with fragile flutes or thin rims, like Forstner bits) is to let them roll around in a drawer with other loose bits.

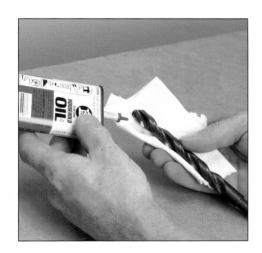

Light coat of oil Most drill bits will benefit from an occasional swipe of oil. I apply a couple drops of light machine oil to a clean soft rag and then wipe the bit with the rag, forcing it into flutes and crevices. Then I follow this up by wiping it dry with another soft rag.

If you've got a separate set of twist bits that you use primarily for nonwood tasks, go ahead and leave a slightly thicker coat of oil on the bit—it'll help prevent rust.

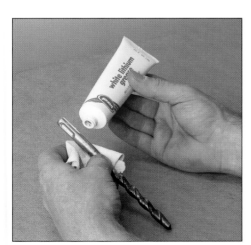

SDS bits If you're using a hammer drill or rotary hammer drill that takes SDS-type drill bits, you should apply a small amount of grease directly to the shank of the drill bit near the notches before inserting it into the chuck.

Almost any kind of medium-viscosity grease will do; I find that white lithium grease works well for this.

MAINTAINING BITS

Just like any quality tool, a set of quality drill bits can be a large investment. It just makes sense to keep them in tip-top shape. Most homeowners think this means keeping them sharp, and this is a big part of it (*see page 111* for more on this). But maintaining your bits also means keeping them burr-free, clean, and lubricated.

As with your drill or drill press, it's a good idea to inspect a drill bit before each use. Take the time to examine the cutting edges, the flutes (if any), and the shank. Burrs of any kind will prevent the bit from operating at peak efficiency, and a burr left unattended can shorten the life span of the bit or possibly cause an accident.

A visual inspection takes only a few seconds and is a habit well worth developing. After a while, you won't even realize that you're doing it until you detect problems—many of which have simple solutions (*see below*).

1 Deburr shanks Burrs are usually caused by not tightening the drill chuck sufficiently; the loose bit spins inside the chuck, scraping along the jaws, raising metal slivers or burrs.

If you discover a burr on the shank of a bit, it's easy to remove. Simply cradle the bit in a piece of emery cloth or wet/dry silicon-carbide sandpaper and, while applying pressure to the shank, spin the bit until the burr is removed. Severe burrs can be filed off with a small mill file, as shown.

2 Honing flutes With use, the flutes of a twist bit or brad-point bit can get nicked and dinged. When this occurs, it often creates burrs inside the flutes, which can impede chip ejection.

If you find any burrs while inspecting your bits, or if you notice the bit is clogging often, you can remove them with a diamond hone (*see page 113*), a small rattail file, or a slip stone. A few strokes is all it takes.

It has always amazed me that even experienced do-it-yourselfers and avid woodworkers tend to shy away from sharpening drill bits. When their bit gets dull, they'll usually apply more pressure to the bit, causing increased friction and heat. This in turn accelerates dulling the bit and can quickly overheat the drill motor.

The other common dull-bit solution is to throw it out and buy a new one. Granted, twist drill bits are inexpensive, but it's not always convenient to run out to the store for a new bit.

Instead of using either of these two options, let's look at a more economical (and satisfying) solution: sharpening the bit. I'll take you through sharpening four of the most common drill bits: twist bits, brad-point bits, spade bits, and Forstner bits.

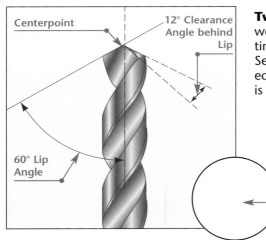

Centerpoint

12° Clearance
Angle behind
Lip

60° Lip
Angle

Thickness of Web
Should Be No Less
Than ⅛" the
Diameter of the Bit

Twist bits Although a twist bit only has two cutting edges to worry about, these edges have a complex grind. First, a 60° cutting edge angle provides best all-around drilling performance. Second, a 12° clearance angle is ground behind the cutting edge to allow the bit to enter the material easily. The challenge is that these two angles need to be ground at the same time.

ANGLE GAUGE

A simple and effective way to check the angle and concentricity of your twist bits is to use a shop-made angle gauge. The gauge is nothing more than a thin piece of hardwood, hardboard, or aluminum with a 60° notch in it. Graduations marked ⅟₃₂" apart provide an accurate reference.

In use, the bit is placed against the gauge; both edges should be perfectly parallel to the gauge. Then rotate the bit to check the position of the drill point against the graduations on the gauge.

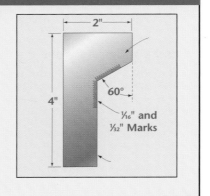

2"

4"

60°

⅟₁₆" and
⅟₃₂" Marks

Twist-bit sharpeners There are a number of ways to tackle the complex grind of a twist bit. The simplest (but most expensive) is to purchase an electric twist-bit sharpener.

These operate much like an electric pencil sharpener: Just insert the bit and rotate it to sharpen the edge. There are also commercially available sharpening jigs out there to fit your grinder. Try to find one that allows you to easily adjust for both height and angle.

Freehand on a grinder For the more economically challenged, you can try sharpening a bit freehand on a grinder. This is cost-effective if you've got a steady hand and are willing to scrap a few bits as you gain proficiency.

To sharpen a twist bit freehand, place the bit on the grinding rest so that the cutting edge of the bit is parallel to the face of the grinding wheel. Then slowly roll the bit to form a slightly convex surface. Continue on the other cutting edge, checking the angles constantly (*see the sidebar on page 111*).

Spade bits The scraping edges on a spade bit are sharpened at a 10° angle. To ensure accuracy, it's best to cut a scrap block at a 10° angle to support the file. With the bit and scrap block clamped firmly in place, take a few upward strokes with the file.

What you're looking for here is an edge that's uniformly smooth and shiny. It's imperative that you leave the outside edges of the bit alone. Any metal removed will alter the diameter of the hole that you're drilling.

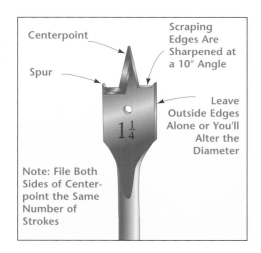

Centerpoint

Scraping Edges Are Sharpened at a 10° Angle

Spur

Leave Outside Edges Alone or You'll Alter the Diameter

1 1/4

Note: File Both Sides of Centerpoint the Same Number of Strokes

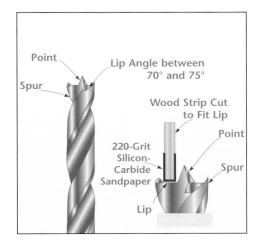

Point
Spur
Lip Angle between 70° and 75°
Wood Strip Cut to Fit Lip
Point
220-Grit Silicon-Carbide Sandpaper
Spur
Lip

Brad-point bits There are three edges that need attention on a brad-point bit: the lip, the spur, and the point. Here again, the easiest way to sharpen the lip is to use an angled scrap block. The only difference is that you'll need to cut a V-groove for the bit. A few strokes on each lip is all it should take.

Next, the spur can be sharpened by tilting the file at an angle. Go lightly here and take smooth, even strokes. Finally, sharpen the point, taking the same number of strokes on each edge.

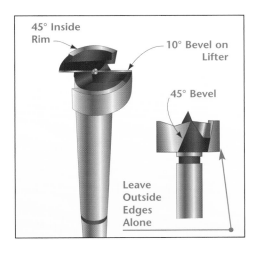

45° Inside Rim
10° Bevel on Lifter
45° Bevel
Leave Outside Edges Alone

Forstner bits Sharpening a Forstner bit is a challenge mainly because it's difficult to access the various cutting surfaces. There are three areas to concentrate on: the lifter, the lifter bevel, and the rim.

The best tools for all of these sharpening jobs are diamond hones (*see the sidebar below*). Start by placing a hone flat against the lifter, then stroke it against the surface. Next sharpen the lifter bevel by stroking the hone against the edge. Finally, sharpen the inside bevel on the rim with a hone or small slip stone.

DIAMOND HONES

Ever since I discovered diamond hones a number of years ago, they've become a permanent fixture in my tool repair box. Diamond hones are basically plastic files impregnated with diamond particles. They come in a variety of shapes, sizes, and grits (typically, coarse, medium, and fine). My favorites are a set of paddles. I use them for everything from reshaping a bit to honing a razor-sharp edge on cutting tools. Diamond hones are available through most mail-order tool companies and can be found at many hardware and home centers in the sharpening-supplies aisle. I also keep a set of needle files and small triangular files close at hand for sharpening in tight places.

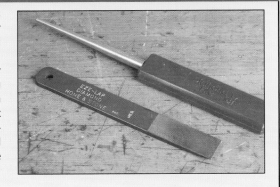

CHAPTER 7

TROUBLESHOOTING COMMON PROBLEMS

Even with well-maintained drills and accessories, you're likely to encounter an assortment of problems as you drill that are as diverse as the wide variety of materials you'll be drilling into.

The most common problems you're likely to encounter when drilling with either a portable drill or a drill press fall into five general categories: inaccuracy where the bit wanders or the hole isn't straight, burning, the holes aren't clean, the drill bogs down or is sluggish, or the bit catches or stops.

Inaccuracy when drilling shows up in two ways: The bit wanders,

or the hole isn't straight. Both of these can be avoided with a little extra preparation time. No rocket science here, just some common sense and a dose of patience (*see pages 115 and 116–117*).

Burning or overheating, whether in wood, plastic, glass, metal, or masonry, is the result of excessive friction—most often caused by using a dull or dirty bit, too high a speed, or too aggressive a feed rate. *See pages 118–119* for the solutions to these problems.

Rough or elongated holes are commonly caused by bits that are dull or dirty, jammed flutes, incorrect speed and feed rates,

and not keeping the drill bit steady. Here again, the solution to most of these problems is fairly straightforward; *see pages 120–121.*

Finally a drill or drill press that bogs down, catches, or stops is often due to insufficient power (often dead or low batteries on a cordless drill, worn-out brushes on a corded drill, or insufficient belt tension on a drill press). As you might suspect, speed and feed rate also come into play here, as well as bits. Fortunately, the solutions to most of these are simple; *see pages 122–125.*

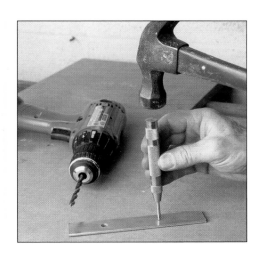

Lack of centerpunch The number one cause of a wandering bit is the lack of a centerpunch. Without a starting point, the tip of the bit will skitter across the surface, leaving designs like a kid's Spirograph.

All it takes to prevent this is to create a starting point with a centerpunch or an awl and a sharp tap of a hammer.

Use guides Another reason a bit wanders has to do with the material you're drilling into. Say, for instance, you need to drill into the end grain of a board. The very nature of the wood will cause problems.

The rings you see on the end of the board are growth rings, or layers of earlywood and latewood. Since the earlywood (light rings) is softer than the latewood (dark rings), the bit will naturally follow the path of least resistance. The solution is to use a guide block to support the bit and guide it on the correct path.

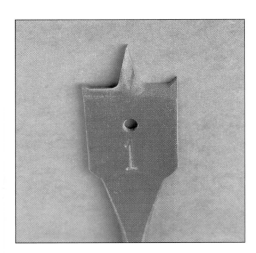

Asymmetrical bit Still another cause of a wandering bit is that the tip, spur, or lead point on the bit is not concentric. When the tip is shaped such that one side of the tip is longer than the other (like the bit shown), it has a tendency to lead or pull the bit in one direction.

The remedy for this is to reshape the tip. As you sharpen the drill point, the critical thing is to make sure to use the same number of strokes on each edge.

Drills & Drill Presses

CROOKED HOLES

Problem The other side to the inaccuracy problem manifests itself as holes that aren't straight. You'll see this as holes that end up drilled at an angle instead of straight, or even as slightly curved or bowed holes when boring with a spade bit.

The cause of both of these problems is a lack of support while the bit enters and continues into the material. Fortunately, support is easy to provide; *see below.*

Use a guide block The type of support you choose will depend on the bit you're using. If you're drilling with a twist bit or brad-point bit, a guide block is just the ticket.

This can be a commercially made guide block or one of the shop-made supports shown on *pages 80–81*. Whenever possible, clamp the guide block to the workpiece for even better support and bit control.

Use a sliding drill guide If the bit you're drilling with is a spade bit or a Forstner bit, the best way to provide support is to use a sliding drill guide.

A sliding drill guide will prevent the bit from wandering off its true path as it drills the hole. Here again, you can purchase a sliding drill guide (like the one shown), or build the shop-made version shown on *pages 84–85*.

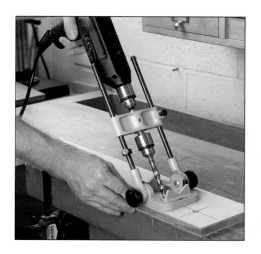

Troubleshooting Common Problems

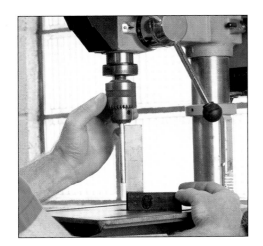

Drill press: Check the table Crooked holes on a drill press are often caused by a table that's not perpendicular to the bit.

Check this by inserting a large-diameter twist bit or metal rod in the chuck. Then raise the table to meet the tip of the bit, and butt a small try square up against the bit. Loosen the table-tilt nut or bolt just enough to be able to adjust the table for square, then retighten it.

Drill press: Check the quill Another common cause of crooked and wandering holes on a drill press is a quill that's out of alignment.

You can check the quill for "runout"—a condition where the quill doesn't rotate straight and true—using either a dial indicator or a scrap of wood and a feeler gauge; *see page 103* for more on this. Runout can also be caused by a chuck that's off-center; *see page 104.*

Check the bit Finally, if your bit is bent and no longer straight, it will cause a number of problems, including crooked holes and wandering. This is particularly noticeable as you start to drill and the tip of the bit wobbles around.

Check a bit by rolling it on a known flat surface, like a drill press table or other stationary power tool. Unfortunately, there really is no solution to this except to throw out the bent bit and get a new one. You can try to rebend the shank; but believe me, it's a lot less frustrating to purchase a new bit.

BURNING

It has always seemed strange to me that a bit that burns wood can lure the average drill user into continuing to use it even after smoke appears. But it's true. Time and time again, I've witnessed someone creating a hole by friction instead of drilling it cleanly.

Some bits, like Forstner bits and holes saws, are especially prone to burning. They require special techniques to keep them from spewing out clouds of smoke; *see below and page 119.*

Even the species of wood has an impact on the likelihood of burning. Cherry has a very well deserved reputation for burning. I've managed to discolor edges of cherry with a brand new ultra-sharp drill bit and the perfect feed rate—that's how finicky it is. But that's not to say that you can't reduce the problem by using the correct drilling techniques (*see below*).

One of the simplest things you can do is to keep your bits clean. Resin, sap, or dirt that has built up on the cutting edges can create unwanted friction, resulting in burning. *See page 108* for more on cleaning bits.

Problem Burning when drilling into wood is most often caused by dull bits, too high a speed, or too fast a feed rate. Often overlooked, another cause of burning is when the flutes of the bit get jammed with shavings—you're basically rubbing wood against wood.

A good habit to get into is to pull the bit out periodically to clear the shavings (preferably with a small brass brush, as the flutes are sharp and the shavings can be hot).

Use a lower speed If your bit is dull or you're using too high a speed, you're likely to experience burning. In the first case, sharpen or replace the bit. In the second, slow down the speed.

On a portable drill, select the lower speed range or lighten up on the trigger finger. With a drill press, change the belt for a slower speed; if you're drilling with a Forstner bit, use the slowest speed possible—these bits were originally designed for use in a hand-powered bit brace.

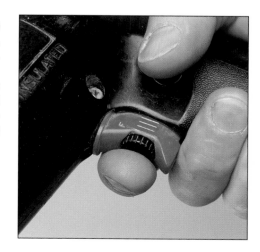

Decrease your feed rate Burning also can occur if your feed rate is too high. The solution is simple: Back off on the pressure and let the bit do the work.

When working with twist bits or brad-point bits, backing off on feed rate also allows the flutes to excavate shavings more effectively from the hole, which also reduces the risk of burning. Remember, if the drill or drill press bogs down and your bit is sharp, it's a sure sign that your feed rate is too aggressive.

Use the proper lubricant If you're drilling into metal, glass, plastic, or masonry, burning or overheating a bit can be caused by not using a lubricant or by using an incorrect lubricant. *See the chart on page 50 for more on this.*

Just as with wood, burning in metal, glass, plastic, and masonry can also be caused by drilling at an incorrect speed and using too aggressive of a feed rate.

Sawdust relief One specific drill accessory, the hole saw, has a really nasty reputation for burning. The reason for this has to do with the inability of the small teeth on the hole saw to handle the dust they produce.

The tiny gullets of the teeth can hold only so much dust, and then you're back to the wood-rubbing-on-wood thing again. The solution? Provide a place for the dust to go. Just drill a dust-relief hole near the inside edge of the perimeter of the hole.

Drills & Drill Presses

HOLES ARE ROUGH

Problem Tear-out and chip-out in wood, rough edges like those shown, scored or lopsided holes drilled in metal, and melted plastic are all signs of holes that aren't being drilled cleanly.

The top cause in all these cases is a dull bit (*see below*). Other common causes are jammed flutes, incorrect speed and feed rate, and not providing support behind the workpiece for when the drill bit breaks through.

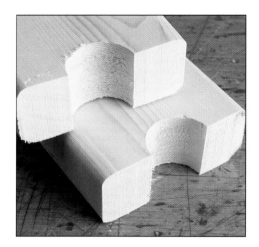

Sharpen the bit If your bit isn't sharp, it can't cleanly cut into wood, metal, or plastic. Instead of cutting, it ends up scraping—a really ineffective way to make a hole.

This creates a domino effect. The scraping dulls the bit even more, so you press harder. This creates more friction, which can cause the bit to lose its temper (hardening). The softer metal dulls even more quickly, and so on. The solution: Sharpen the bit (*see pages 111–113 for more on this*).

Flutes are jammed Rough holes can also be caused by chips that get stuck in the flutes of the bit; they rub against the walls of the hole, scoring it and tearing out other chips.

To prevent this, periodically pull out the bit and clear the shavings. If you notice that the flutes jam often, check to make sure there aren't any nicks or burrs that can hinder ejection (*see page 110*). If you do pull out a bit that's jammed with shavings, don't use your fingers to clear the flutes—the flutes are razor-sharp, and the shavings are darn hot. Keep a small brass brush on hand for occasions like this.

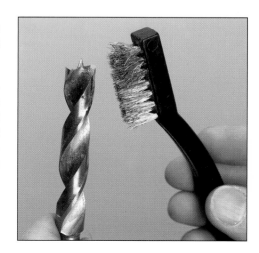

Troubleshooting Common Problems

Speed and feed rate As usual, drill speed and feed rate also come into play when rough holes are encountered. If you're forcing the bit into the material, you're not giving the bit the time it needs to cut cleanly; you're also not allowing the chips to eject completely.

Slow down and/or back off the pressure; some bits (such as Forstner bits and hole saws) operate best at very slow speeds and firm but steady feed rates.

Use a backer board When drilling into any material, you can prevent chip-out as the bit exits the hole by simply providing support behind your workpiece.

With a portable drill, clamp a scrap piece of wood behind the workpiece. On the drill press, simply slip a scrap of wood under the workpiece. In most cases, you'll get the best results by clamping the workpiece and scrap block to the drill press table.

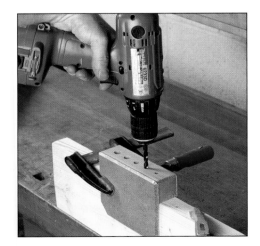

Steady drill Rough holes can also be caused by a bit that moves around—in effect, you're reaming out the hole.

To prevent this from happening, use some form of bit support, such as the shop-made 2×4 saddle shown; *see page 81* for instructions on how to make this. Or you can use a simple guide block like the one shown on *page 65,* or purchase a commercially made drill guide.

DRILL BOGS DOWN

Cordless drill: Battery is weak When using a portable drill, you may occasionally notice that the drill bogs down or is sluggish. This is a prime indicator that something is wrong.

The most obvious thing to check is your drill's power source. If you're using a cordless drill, try a fresh battery or recharge to see whether this eliminates the problem. If not, it's likely that you're using the incorrect speed or feed rate (*see below*).

Corded drill: Brushes If you notice that a corded drill is bogging down a lot, it's a good time to check your brushes; *see page 98* for more on this. Some drill manufacturers, like the one whose drill is shown here, make this a quick and easy task. With others, you'll need to remove an access panel or even split the case.

If the brushes have less than ¼" of carbon left, they need to be replaced. Check the spring tension as well—tired springs won't provide good contact, and the drill will operate sluggishly.

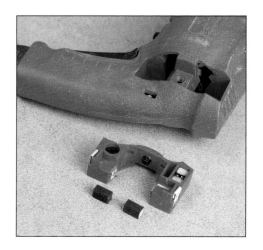

Portable drill: Speed and feed Again, speed and feed rate are a factor. Listen to the drill. If you're sure your bit is sharp and the speed is correct, ease off on the feed rate.

If the drill still bogs down, you may be asking too much from it. Remember that the lower speed range provides much higher torque than the high-speed setting. Use the low range when driving screws and drilling into very hard materials.

Drill press: Slipping belt If your drill press bogs down while drilling a hole, especially a big one, the problem may be that the belt is slipping. If it is, you'll often hear it: a high-pitched squeal, much like a slipping belt on a car.

To correct this, loosen the tension-lock knob and push the motor farther away from the spindle (*see page 105* for more on this). When the tension is correct, tighten the tension knob.

Drill press: Incorrect speed Another reason your drill press may be bogging down is that you're using the incorrect speed for the material. Check the chart on *page 46* to match the speed to the material.

First unplug the drill press and change speeds by loosening the tension knob and lifting up the motor/pulley cover. Slip the belt off one of the step pulleys, and reposition it to the desired speed. Apply the correct amount of tension, lock the tension knob, and close the cover.

Drill press: Bit is dull The final reason a drill press will bog down is because the drill bit is dull. I mention this last because the heavy-duty motor in most drill presses can usually drill a hole even with a dull bit. But if the bit is super-dull, it may slow down the motor.

Check the bit and sharpen it as necessary. *See pages 111–113* for details on sharpening common bits.

DRILL CATCHES OR STOPS

There are a couple of reasons for a drill to catch or stop. Number one is that you've hit something that the bit couldn't handle.

In wood, this may be a knot or some particularly dense or twisted grain. In metal (especially sheet metal), the bit may be catching the edge of the thin metal as it breaks through. In masonry, you could have hit a tough piece of aggregate.

Whenever this happens, reverse the drill (if that's possible) and back the bit out of the hole. Once the bit has been extracted from the hole, the best thing to do is take a step back and find what caused the stoppage.

It may be as simple as a dead battery, a bad brush, a disconnected extension cord, or a blown breaker or fuse. Assuming that these are all fine, reinsert the bit in the hole and give it another try. Take it easy and apply gentle pressure. Be ready to release the power switch instantly.

Drill a pilot hole If your drill bit persists in catching, try switching to a smaller bit to break through the material. This does two things.

First, it removes some of the material, making it easier for the larger bit to tackle. Second, it provides a channel for the larger bit to follow to help guide the bit though the tough material.

Sheet metal When drilling into thin sheet metal, there's a natural tendency for the bit to catch and jam as the bit breaks through the opposite side. You'll notice this most often when you're drilling large holes.

The best way to cope with this is always to support the opposite side. A quick solution is to clamp the metal tightly between two scraps of wood, then drill through the wood and metal sandwich.

Drill press: Belt isn't tensioned properly You're drilling into a workpiece on the drill press, and the bit catches and stops. The belt slips, and the motor continues to turn. Although this is a nuisance, it's also a safety feature: It prevents the bit from damaging or grabbing the workpiece.

The first thing to do is figure out what caused the bit to catch—it's often a knot or otherwise hard material. If after you back the bit out and try again the bit continues to catch and stop, try increasing the belt tension (*see page 105*).

Drill press: Pulley is loose The other reason a bit in a drill press will catch and stop is that one of the pulleys may be loose on the motor shaft or spindle.

You can check this by unplugging the machine and lifting the motor/pulley cover. Then try to wiggle each step pulley in turn. If either one is loose, tighten the respective setscrew. Also make sure that the bit is secure in the chuck and not simply spinning inside it.

GLOSSARY

Backer board – a scrap of wood clamped behind or inserted underneath a workpiece to support the workpiece and prevent tear-out when the bit breaks through the other side.

Bench-top drill press – a type of drill press intended for use on a bench or other table; often as powerful as its larger floor-model cousins.

Brad-point bit – a drill bit (often referred to as a dowel bit) that's similar to a twist bit but has a different tip—a sharp lead point and a pair of cutting spurs—which makes it ideal for cutting clean holes in wood.

Buffing wheel – a cloth or rag wheel that accepts an arbor to fit either the portable drill or drill press; used for buffing and polishing—with or without buffing compounds.

Circle cutter – a specialty bit designed for use only on a drill press to cut large holes; uses a pilot bit and one or two cutters to score the perimeter of the hole.

Cordless drill – a battery-powered drill that offers portability since it has no cord. Battery-charging systems recharge batteries in as little as 15 minutes.

Counterbore – to enlarge the upper part of a hole to accept and allow the head of a screw or bolt to be recessed below the surface.

Countersink – a cone–shaped chamfer cut in material to accept the tapered head of a screw so it lies flush or just below the surface.

Chuck – the adjustable jaws on a drill or drill press for holding bits or other drilling accessories; may be either keyed or keyless.

Chuck key – a geared key used to tighten the jaws on a keyed chuck.

Column – a sturdy hollow-metal rod that attaches to the base of the drill press and supports the drill head; an adjustable table also rides on the column to raise or lower the workpiece.

Depth stop – an adjustable stop, often a metal rod or collar that comes to rest against the workpiece when the drill bit has reached the required depth; or an adjustable stop rod or rotating cam that stops the quill on a drill press at a specified depth.

Disk sander – a flat, flexible rubber disk with an arbor for use with a portable drill; self-adhesive sandpaper in a variety of grits attach to the rubber disk for aggressive sanding jobs.

Doweling jig – a jig that positions and supports a drill bit to drill holes for wood dowels.

Drill index – a box, usually metal, with pivoting pre-drilled holders inside to store and organize twist bits.

Drill press – a stationary power tool designed to drill holes and a variety of other tasks; there are three major types: bench-top, floor-model, and radial.

Driver/drill – a type of portable cordless drill that both drills holes and can be used to drive screws;

usually has two speeds: a low speed with high torque for driving screws, and a high-speed, low-torque setting for drilling holes.

Drum sander – a cylindrical drum with an arbor to fit a chuck that accepts sanding sleeves in a variety of grits; can be used in either a portable drill or a drill press to sand and shape curved wood parts.

Feed rate – the rate at which a drill bit or accessory is pushed or forced into a workpiece.

Fence – an adjustable guide usually made of wood that's clamped to a drill press table to reliably position a workpiece for drilling.

Flap sander – a sanding accessory (sometimes referred to as a flap-wheel) made up of hundreds of cloth abrasive sheets attached to a core; useful for sanding hard-to-reach convex and concave surfaces.

Floor-model drill press – a type of large drill press that rests on the floor. The extra-long column allows for drilling into long or large workpieces.

Flutes – the twisted grooves or channels in a drill bit intended to excavate chips for a hole being bored.

Forstner bit – a specialty bit that drills clean, flat-bottomed holes; the continuous rim allows the bit to drill overlapping and partial holes with ease.

Grinding wheel – an abrasive wheel with an arbor to fit either a portable drill or a drill press; used to sharpen and shape metal.

Hammer drill – a type of impact drill that delivers several hundred blows per second to the drill bit to help the bit break up masonry when drilling into stone or brick.

Hold-down – an adjustable two-fingered bracket that's part of a mortising jig and is used to hold down the workpiece as the mortising bit is repeatedly inserted and removed from the workpiece.

Hole saw – a cup-shaped saw blade with a centered drill bit used to cut large holes in various materials.

Impact drill – a type of drill, either corded or cordless, that combines percussion with drilling action to cut into masonry; the two main types are hammer drills and rotary hammer drills.

Keyless chuck – a three-jaw chuck that can be fitted on a portable drill or drill press that doesn't require a chuck key to tighten the jaws.

Masonry bits – twist bits with carbide tips brazed onto the ends, designed to cut into tough masonry materials like brick, stone, and concrete.

Mortising attachment – a special attachment that bolts to the drill press and allows you to cut square holes or mortises; uses special two-piece mortising bits consisting of a rotating bit that fits inside a hollow-square chisel.

Multi-spur bit – a derivation of the Forstner bit, which instead of a continuous rim has a set of jagged teeth. The gullets between the teeth whisk away chips so the bit can be run at higher speeds.

Non-impact drill – any portable drill that does not offer the option of mixing percussion with drilling as an impact drill does.

Pilot bit – a drill, countersink, and counterbore bit wrapped into one; designed to drill the three-step clearance hole required by tapered woodscrews.

Plug cutter – a specialty bit designed for the drill press to cut cylindrical plugs of wood to hide the heads of counterbored screws.

Quill – the sleeve surrounding the spindle of a drill press; how deep a hole a drill press can bore in one pass depends on the amount that the quill can be raised and lowered.

Rack-and-pinion – a two-part system used on a drill press to raise and lower the table; a toothed bar attached to the column accepts a geared wheel that's part of the table height adjustment mechanism.

Radial drill press - A radial drill press has a much larger throat capacity (typically around 36") than other drill presses; its head is attached to a horizontal ram column that can be rotated either to the left or to the right as well as in and out.

Runout – a condition on a drill press where the quill doesn't run straight and true: The bit will wobble, and starting a hole at a precise point will be almost impossible.

SDS bit – a type of masonry bit; SDS stands for "Steck-Dreh-Sitzt." Roughly translated from German this means "insert-turn-locked-in." Indentations in the shank provide the chuck with a firm purchase to guarantee that the bit won't come flying out in use.

Spade bit – an inexpensive, paddle-shaped bit with a long lead point that's designed for power drilling of large-diameter holes, typically ⅜" and larger.

Spindle – the vertical rotating shaft of a drill press; a three-jaw chuck is pressed on the spindle to accept and grip bits.

Spline bit – a type of masonry bit with lengthwise slots in the shank to offer much more surface area for the transmission of power, while at the same time affording a rock-solid grip.

Step pulley – a motor pulley with different levels or steps that allow you a range of speed choices; change speeds by moving a V-belt to the different diameter steps.

Stop collar – a metal sleeve that slips over a drill bit and is locked in place with a setscrew to set the drilling depth.

Swarf – the metal filings or shavings resulting from drilling a hole.

Tear-out – the tendency of a bit to tear the fibers of the wood it is cutting, leaving ragged edges on the workpiece, especially when a bit exits a hole.

Trigger lock – a button on the side of a portable drill that when depressed allows the motor to continue running without holding down the trigger.

Twist bit – a drill bit made with a pair of helical flutes that clear the waste from the holes as the bit bores into the material. The flutes terminate in two cutting edges to form a pointed tip.

Wire wheel or brush – a wire wheel fitted with an arbor or a wire brush can be used to quickly remove rust and old paint from metal.

INDEX